Common Seaweeds of India

Common Seaweeds of India

Dinabandhu Sahoo

I.K. International Publishing House Pvt. Ltd.

NEW DELHI • BANGALORE

Published by

I.K. International Publishing House Pvt. Ltd.
S-25, Green Park Extension
Uphaar Cinema Market
New Delhi – 110 016 (India)
E-mail: info@ikinternational.com

ISBN 978-81-907770-6-3

Published by Krishan Makhijani for I.K. International Publishing House Pvt. Ltd., S-25, Green Park Extension, Uphaar Cinema Market, New Delhi – 110 016 and Printed by Rekha Printers Pvt. Ltd., Okhla Industrial Area, Phase II, New Delhi – 110 020.

Preface

Under the vast expanse of tropical waters of Bay of Bengal and Arabian Sea there lies huge resources of marine life which consist of beautiful corals, seagrasses, fish, marine mammals, marine algae etc. While common people are aware of corals, whales many of them are less informed about the presence of a group of colorful plants called seaweeds. These macromarine algae have been playing a very important role not only in our day to day lives but also in the marine ecosystem. To make the common people, students, policymakers, entrepreneurs and researchers aware about the seaweeds and their uses, I wrote my first book "Farming the Ocean: Seaweeds Cultivation and Utilization" in the year 2000 which become very popular among people from various walks of life. This book also encouraged many students to study Phycology (study of algae) and enthused some young entrepreneurs to set up seaweed businesses in India. My second book "Seaweeds of Indian coast" Sahoo *et al.* 2001 was one of the most difficult jobs where a checklist of 770 species was compiled from different parts of the Indian coast.

Then I was requested by many of my masters and research students to prepare field guide that would be helpful in identification of seaweeds. While searching the literature on marine algae field guides I could only find the monumental work by Srinivasan (1969, 1973) in Phycologia Indica containing 50 colorful hand drawings. While making the outline of this book I thought it would be appropriate to concentrate only on common seaweeds of India rather than all the 770 species. Hence the title of the book "Common seaweeds of India".

This book contains original pictures taken by our group during the field trips undertaken to various parts of the Indian coast including Andaman & Nicobar Islands as well as Lakshadweep group of Islands.

I hope this book will be useful for students, researchers, teachers, nature lovers, scuba divers and people from various walks of life.

Delhi, India
December 1, 2009

Dinabandhu Sahoo

Acknowledgements

I am extremely grateful to my supervisor Late Professor M.R. Vijayaraghavan who guided my research career. I also thank my close Japanese friend Professor Shunji Hosaka, Chuo University, Japan for his constant encouragement. I thank Dr. I. A. Levine from University of Southern Maine, USA who is presently a visiting New Century Fulbright Scholar in my laboratory at Delhi University for his help.

This book would not have been possible without unconditional and tireless efforts of my research students Dr. Pooja Baweja, Mr. M. Arunjit Singh, Ms. Elangbam Geetanjali, Mr. Savindra Kumar and Ms. Salam Sonia Devi. At the same time I received enormous help from my other research students Gaurav Kumar, Nitin Pipralia, Arvind K. Bhardwaj, Priyanka Verma and Vivek Chopra. The contribution of my former students Dr. Nivedita, Dr. Debasish, Dr. Sunita Singh, Dr. Saswati Nayak, Ms. Mrinmala Devi, Vipin Dhama, Neetu Kushwah, Babita Kumari cannot be ignored.

I thank my wife Dr. Meenakshi Munshi, son Chaitanya Sahoo and daughter Aanchal Sahoo for their constant support.

I thank the publisher IK International for agreeing to publish this book in a timely manner.

Dinabandhu Sahoo

Contents

Introduction

Marine algae, popularly known as seaweeds, are one of the most beautiful groups of photosynthetic organisms which grow under the ocean's blue water. For a common man these plants are of little use and very few people know that these seaweeds are used as food, feed, fodder, fertilizer, and medicines since ancient times. The extracts from some of these seaweeds are used in ice cream, tooth paste, tomato ketchup, dentistry, microbiology, biotechnology, textile printing, chocolate, meat pressing, and several other industries. Today seaweeds and their products are a multibillion dollar industry and the demand for these products will increase in the future. Various pharmaceutical companies are now prospecting for new biomolecules from seaweeds.

There are three main groups of seaweeds which can be identified on the basis of their color as Green (Chlorophyta), Brown (Phaeophyta), and Red (Rhodophyta) algae. The difficulty is the correct identification of these seaweeds. An interesting aspect of these enigmatic plants, while immersed in the seas or in fresh conditions these plants exhibit beautiful bright colors but when these plants die or dried up they lose their beauty, shape, size and color. Algae can be preserved on herbarium sheets where they retain a portion of their color and luster but when preserved in liquid preservatives, they lose their color. Therefore, the best way to identify these plants is in the field, *in situ,* where they retain their original characteristics while utilizing a field guide.

Seaweed field guides or manuals are available from other countries, e.g. Common Seaweeds of China (Tseng, 1984), Seaweeds of Hawaii (Magruder ands Hunt, 1979), Marine Plants of the Caribbean (Littler *et al.* 1989), Seaweeds of Japan (Tokuda *et al* . 1994), The Common Marine Plants of Southern Vietnam (Isao *et al.* 2005), and Phycologica Indica (Srinivasan 1969, 1973) etc.

India has a long coastline of more than 7000 km and nearly 770 species of seaweeds have been reported from the Indian marine environments (Sahoo *et al.* 2001). During the last 25 years, I have extensively travelled to many remote regions of the Indian coasts including Lakshadweep as well as Andaman & Nicobar island groups. This book examines a diversity of Indian coastal environments from the Gujarat and Lakshadweep Islands off the West Coast and the Andaman & Nicobar Island groups off India's east coast. This volume includes 113 algal species including 31 species of green, 30 species of brown, and 52 species of red seaweeds. Additionally, seaweed cultivation efforts along the Indian coastline are covered. I hope this book will be very useful to people from all walks of life.

SEA COASTS OF INDIA

Map not to scale

Map of India Showing Important Sea Coasts

Okha Coast, Gujarat during low tide

Rocky coast of Shreewardhan, Maharashtra

Tide pools at Shreewardhan Coast

SEA COASTS OF INDIA

Cochin Coast, Kerala

Rocky coast of Thangesari, Kerala

SEA COASTS OF INDIA

Sea coast of Lakshadweep Island

Sea coast of Diu

Aerial view of Andaman Island

Chidia Tapu Coast during low tide, Andaman Island

SEA COASTS OF INDIA

Sea coast at Kanyakumari, Tamil Nadu

SEA COASTS OF INDIA

A view of Pamban Coast, Tamil Nadu

Sea coast of Killakarai, Tamil Nadu

SEA COASTS OF INDIA

Rocky coast of Tiruchendur, Tamil Nadu

Sea coast of Visakhapatnam, Andhra Pradesh

SEA COASTS OF INDIA

Sea beach at Puri, Odisha

Artificial sea mouth at Chilika Lake, Odisha

SEA COASTS OF INDIA

Mangrooves at Sunderbans Coast, West Bengal

NATURE OF SUBSTRATUM

Natural rocky substratum showing extensive growth of seaweeds

Artificial rocky substratum showing extensive growth of seaweeds

NATURE OF SUBSTRATUM

Extensive growth of seaweeds growing in tide pools

Seaweeds growing on corals

Chlorophyceae (Green Algae)

Chlorophyceae (derived from the Greek word ***choloros*** meaning "green" and ***phyton***, meaning "plant") or green algae, are often present in relatively high numbers in tropical seas. The "green algae" is the most diverse group of algae, with more than 7000 species growing in a variety of habitats. The green macro marine algae popularly known as Green seaweeds are found on both sandy and rocky beaches. Many can tolerate low salinity and colonize areas where rivers meet the sea. The green colour is due to the presence of green pigment chlorophyll required for the photosynthesis. Sometimes the excessive growth of some green seaweed species causes "Green tides" in the sea. Some of the most common green seaweeds are *Ulva lactuca* popularly known as sea lettuce, *Caulerpa racemosa* (sea grapes), *Enteromorpha intestinalis* (gut weed or Aonori in Japan) and eaten as food.

CHLOROPHYCEAE

Monostroma oxyspermum (Kützing) Doty

Thalli light green, membranous, form dense tufts, gelatinous in appearance. Plant is 3-10 cm in height, attached to substratum by basal disc. Thalli initially saccate, later split into irregular membranous, with thick intervening walls; cells in surface view angular, round or oval. Thallus arising as a sac from a prostrate discoid portion and rupturing at maturity to form monostromatic blade.

Distribution

Gujarat, Maharashtra, Goa, Karnataka.

Monostroma oxyspermum

Enteromorpha compressa (Linnaeus) Nees

Mature thalli are bright green to dark green. Thalli are 5-12 cm long and 2-5 cm broad and attached to substratum by a small circular holdfast. The basal part of the thallus is round and becomes flattened towards the upper part. The apical part of the thallus is obtusely rounded and often perforated. The thalli are compressed towards the apical end.

Distribution

Andaman and Nicobar Islands, Maharashtra, Tamil Nadu (Madras, Krusadai island, Tuticorin, Mandapam, Mahabalipuram, Pulicut, Mandativu, Gulf of Mannar), Gujarat (Okha), Goa, Andhra Pradesh (Visakhapatnam), Karnataka (Mangalore), Odisha (Chilika Lake).

Enteromorpha compressa

Enteromorpha intestinalis (Linnaeus) Nees

The thalli varied considerably in its external morphology, colour, height and width in different habitats. The thalli grow upto 25-35 cm in length, yellowish-green in colour and turn colourless when exposed to excessive sunlight. The plants are found attached to the substratum by circular disc when young. However as the plants grow longer, they become free-floating many a times covering the entire water surface of that particular area. The filaments are long, contorted several times and often inflated at intervals giving an intestine-like appearance.

Distribution

Lakshadweep Island, Andaman and Nicobar Islands, W. Bengal (Hughli-Howrah, Salt Lake), Goa, Kerala, Gujarat (Gopnath, Saurashtra, Okha), Andhra Pradesh, Odisha (Chilika Lake, Charaiguha), Maharashtra, Karnataka, Tamil Nadu (Krusadai Island, Porto Novo, Muthupet Estuary, Kalpakkam), Pondicherry.

Enteromorpha intestinalis

Enteromorpha flexuosa (Wulfen) J. Agardh.

Mature thalli are yellowish green to light green in colour and grow upto 21 cm. The thallus remains attached to the substratum with the help of a discoid holdfast. The filaments are circular at the lower part and broad in the upper part. Filaments are flexuous towards the upper part. The upper part of the thallus is somewhat obtuse in structure.

Distribution

Lakshadweep Island, Goa, Gujarat (Okha, Dwarka, Saurashtra), Maharashtra (Malwan, Bombay), Tamil Nadu (Madras, Tuticorin, Mandapam, Muthupet Estuary, Tiruchendur), Kerala, Karnataka.

Enteromorpha prolifera (O.F. Müller) J. Agardh

Plants 30-40 cm long, upto 1.5 cm broad, holdfast circular, minute, filaments show tubular at base, compressed above, simple, with a few to many proliferations varying in length, and form several secondary proliferations. In surface view cells look polygonal, linearly arranged in thinner filaments only and not in thicker or older parts. Cells uninucleate, chloroplast filling cell, with one to many pyrenoids.

Distribution

Bengal (Salt Lake), Maharashtra (Bombay) Andhra Pradesh (Waltair Coast), Gujarat (Jalleshwar, Veraval), Tamil Nadu (Tiruchendur), Kerala (Kovalum), Andaman and Nicobar Islands, Lakshadweep Island.

Enteromorpha flexuosa

Enteromorpha prolifera

Ulva fasciata Delile

Young plants are dark green in colour and give a soft touch, older plants become light green, become slimy, generally thalli 12-18 cm long, 1-3 cm broad. A thallus upto 42 cm has also been reported, cells polygonal 15 -18 μm × 7-12 μm in size. Cells at apex are loosely arranged, in middle region cells are quadrangular, closely arranged , larger than cell of apex and base, 10-20 μm × 8-16 μm in size, cells at base are smaller, irregular with round angles, 9 -14 μm × 8-12 μm in size. Plants are attached to the rocks with the help of a small circular holdfast. The thalli are lobed towards the basal part.

Distribution

Lakshadweep Island, Andaman and Nicobar Islands, Tamil Nadu (Krusadai Island, Mandapam, Porto Novo), Goa, Gujarat (Vumani reef near Port Okha, Saurashtra, Veraval), Maharashtra, Andhra Pradesh (Visakhapatnam, Gangavaram), Kerala (Trivandrum, Kovalam), Pondicherry.

Ulva fasciata

Ulva lactuca Linnaeus.

Commonly known as "Sea lettuce". Thalli light to yellowish green, membranous, ovate to roundish, sometimes deeply incised, the margins plane or ruffled, about 30 cm in height, growing on the middle intertidal rocks and rocks pools. Holdfast distinct disc shaped. Cells irregularly arranged 10-20 µm in diameter. Cells in apical and middle part irregular, round in basal region, 10- 25 µm in thickness.

Distribution

Lakshadweep Island, Andaman and Nicobar Islands, Gujarat (Okha, Porbandar, Diu, Baroda, Gulf of Kutch, Vumani reef and Hanumandandi reef near Port Okha, Jalleshwar, Saurashtra), Maharashtra (Bombay, Tarapur Power Station, Malwan), Andhra Pradesh (Visakhapatnam), Tamil Nadu (Krusadai Island, Mandapam, Tuticorin, Porto Novo, Gulf of Mannar, Palk Bay, Tiruchendur), Kerala (Kovalam), Odisha (Chilika Lake), Karnataka (Karwar) Daman and Dui (Devka Coast), Pondicherry.

Ulva lactuca

Ulva reticulata Forsskål

Thallus is net like or reticulate, pale green in colour and grows upto 8-15 cm in length and 10-20 cm in width. They are expanded in sheet, 2 layered in thickness. The plant is attached to a substratum throughout its life by disc-shaped holdfast, thallus smooth, delicate in texture. Cells in surface view are irregularly placed polygonal in shape, each cell contains single parietal chloroplast; often with deeply incised or lobed margin containing a single pyrenoid.

Distribution

Tamil Nadu (Krusadai Island, Mahabalipuram, Palk Bay, Mandapam, Gulf of Mannar, Pamban, Tuticorin, Tiruchendur Mandativu, Muthupet Estuary), Gujarat (Saurashtra), Kerala, Maharashtra (Bombay), Goa, Karnataka, Diu, Bay of Bengal, Andaman and Nicobar Islands.

Ulva reticulata

ULVALES, ULVACEAE

These pictures show extensive growth of *Ulva* sp. in some parts of the Indian coast. The thalli get drifted by wind to the shore as shown in the picture.

Drifted *Ulva* sp. at Mandapam, Tamil Nadu

Chaetomorpha antennina (Bory de Saint-Vincent) Kützing.

Plants are dark green, filamentous, dense, unbranched, stiff, rigid below and flexuous above, 4-10 cm high, attached with hard rocks and well-developed rhizoid which are irregularly ramified. Rhizoid can swell anywhere filled with starch ultimately giving new shoot from them. It grows at high-water mark on all rocky coasts of India. Filament 's end looks colourless because of escape of zoospores or gametes. The alga is heavily invested with *Acrochaetium* sp. and *Erythrocladia* sp. There are a number of records of this species under the name *Chaetomorpha media.*

Distribution

Maharashtra (Bombay, Malwan), Karnataka (Mangalore), Tamil Nadu (Krusadai Island, Palk Bay, Mandapam, Porto Novo, Gulf of Mannar, Tiruchendur), Gujarat (Saurashtra), Kerala (Kovalam) Andhra Pradesh (Visakhapatnam), Andaman and Nicobar Islands, Pondicherry.

Chaetomorpha antennina

Chaetomorpha crassa (C.Agardh) Kützing.

The filaments are fairly green, usually free floating, very long, cylindrical and sometimes slightly swollen. This species form soft, loose clumps of filaments that are unbranched series of cylindrical to barrel shaped cells. Plants drift above the bottom, or entangled with other seaweeds and seagrasses in the intertidal to subtidal zone of shores with moderate to calm wave action.

Distribution

Different parts of Indian coast

Chaetomorpha linum (Müller) Kützing.

Mature filaments are deep green in colour and grow upto 2-3 ft. in length. Maximum length of the thalli is observed during the winter season *i.e.* between December-January. The filaments are unbranched and are attached to the substratum by rhizoid-like structure. The vegetative cells are cylindrical to barrel shaped. The apical cell is round on the tip of the filament. A thick mucilaginous layer covers each filament and the cell wall is fairly thick. The cells in the filament vary in size and are always in an active state of division thereby increasing the length of the filament. Cells are 1-1.5 times longer than broad upto 260 μm in breadth. Basal cell is the longest cell in filament.

Distribution

Lakshadweep Island, Andaman and Nicobar Islands, W. Bengal (Salt Lake), Odisha (Chilika Lake), Karnataka, Maharashtra, Goa, Tamil Nadu (Tuticorin) Gujarat (Jalleshwar, Veraval).

Chaetomorpha crassa

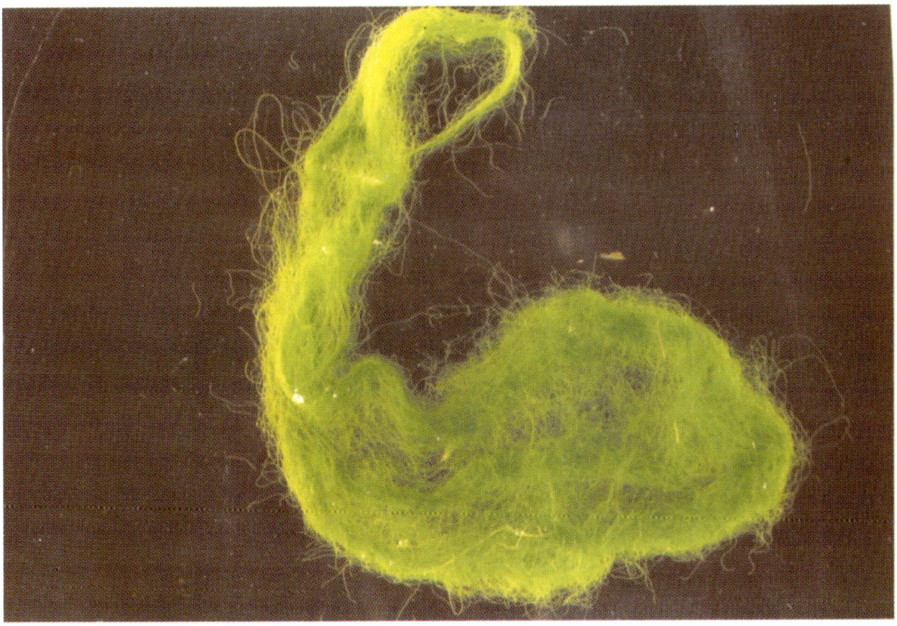

Chaetomorpha linum

Cladophora glomerata (Linnaeus) Kützing

A mature thallus is yellowish - green in colour and grows upto 2-3 m in length. At the early stage, the plants are firmly attached to the substratum and the mature filaments float freely. The filaments are rough to touch. When in attached condition the filaments are highly branched thus gives a more bushy appearance. Branching pseudo-dichotomous and are found upto several orders. The plants grow in association with various seaweeds and seagrasses.

Distribution

Odisha (Chilika Lake), Gujarat (Okha), different parts of the Indian Coast.

Cladophora glomerata

Boodlea composita (Harvey) Brand.

Plants are green, sponge - like masses consisting of intertwined filaments which are soft and loosely arranged. Habitat is affected by local environmental conditions. It can be seasonally dominant forming loose mats upto several depth.

Distribution

Gujurat (Dwarka, Okha, Saurashtra), Maharashtra (Bombay, Konkan), Tamil Nadu, Lakshadweep, Nicobar Island.

Boodlea composita

Boergesenia forbesii (Harvey) Feldmann

Plant is erect, yellowish green in colour looking like small ballons. They grow in clusters and attached to dead coralline or hardy substratum. Thallus broadly clavate,or linear oblong, vesicular, swollen with narrow base and broader upper portions, ballon like, filled with clear sea-water. Number of vesicles vary in size and shape.Vesicles form plenty of cysts inside. This is the only species of *Boergesenia* found in India.

Distribution

Tamil Nadu (Tuticorin, Pamban, Krusadai Island, Palk Bay, Mandapam, Gujarat (Jalleshwar, Veraval), Lakshadweep Island (Minicoy), Andaman & Nicobar Islands.

Boergesenia forbesii

Valonia utricularis (Roth) C. Agardh

Plants are cushion like having a tuft of cylindrical to clavate vesicles. The plants look like hemispherical domes. Plants are green in colour growing upto a height of about 5 cm. The vesicles are profusely branched and interconnected by hapteroid growths of lenticular cells. Plants are commonly found in intertidal rocks and stones in coral reefs, often in tide pools.

Distribution

Gujarat (Dwarka, Okha), Tamil Nadu (Krusadai Island), Nicobar Island.

Valonia utricularis

Caulerpa microphysa (Weber- van Bosse) J.Feldmann

Plants are small, forming dense clusters on corals and rocks grow upto, 15 cm tall. The thalli are bright dark green in colour, terete, naked stolon less than 1 mm in diameter, bearing descending branches with branched, filiform rhizoids at the ends. The ascending branches on the upper side of the stolon shortly stipitate, simple or occasionally branched, with terete rachis bearing crowded, spherical ramelli, 1-1.5 mm in diameter, supported by stalks, 0.5 mm long, constricted at the apical end. The plants have slender creeping rhizomes.

Distribution

Tamil Nadu (Mandapam, Gulf of Mannar, Palk Bay), Nicobar Island.

Caulerpa microphysa

Caulerpa fastigiata Montagne

This species has filamentous and cylindrical branches that divide dichotomously or alternately. The runners creep along the substrate. Plants are found on sandy rocks and gravel in the mid to lower intertidal zone of calm shores, such as on inner reef flats.

Distribution

Maharashtra (Bombay, Konkan), Gujarat (Okha), Andhra Pradesh (Visakhapatnam), Tamil Nadu (Mandapam, Madras), Lakshadweep Island.

Caulerpa racemosa (Forsskål) J. Agardh

Thalli are pale green, 5-6 cm tall. Plants in larger or smaller group possess rhizome with creeping, cylindrical and highly branched. Rhizoids arise from the lower parts of prostrate rhizome. Crowded with bud-like branchlets in spherical shape, sometimes oval not constricted at the base. They are found growing on sandy, rocky or silt-covered substrate attached by numerous rhizoids arising from the creeping stolon. It was observed that the thalli were rarely associated with other macroalgae.

Distribution

Maharashtra (Bombay, Malwan), Gujarat (Dwarka, Okha, Saurashtra, Veraval), Tamil Nadu (Palk Bay, Pamban, Mandapam, Tuticorin, Krusadai Island, (Gulf of Mannar), Kerala (Kovalam), Andhra Pradesh (Visakhapatnam), Goa, Karnataka, Lakshadweep Island, Andaman and Nicobar Islands.

Caulerpa fastigiata

Caulerpa recemosa

Caulerpa peltata Lamouroux

Plants are small, generally bright green in colour. The stolons freely forked, giving off rhizoids-bearing branches below and foliar branches above, erect axis varying between 1-10 cm in length with numerous short branchlets which radiate from all directions. Each branchlet terminates in a peltate assimilatory disc.

Distribution

Tamil Nadu (Krusadai Island, Pamban, Kovalam, Mandapam), Gujarat (Okha, Saurashtra, Jalleshwar), Goa, Kerala, Maharashtra, Andaman and Nicobar Islands.

Caulerpa sertularoides (S.Gmelin) Howe

Plants look extremely beautiful with its bright green dark colour. Thalli have prostrate branched cylindrical rhizome-like stolons creeping on substrata, anchored by well branched rhizoids below, forming large colonies. Stolons naked with a number of upright, delicate and flexible cylindrical axis with lateral outgrowths the assimilators, on the upper side. Assimilators are flattened, feather like, 10-15 cm long and 13-22 mm broad.

Distribution

Maharashtra (Bombay), Tamil Nadu (Krusadai Island, Mandapam, Gulf of Mannar, Palk Bay, Tiruchendur), Gujarat (Okha), Goa, Karnataka, Andaman & Nicobar Islands.

Caulerpa peltata

Caulerpa sertularoides

Caulerpa scalpelliformis (R. Brown ex Turner) C. Agardh

Thallus is dark green with extensive stoloniferous prostrate axes bearing upright frond upto 10 cm in height. They are attached to the substratum by fine branches. Upright fronds are terete near the base, bearing closely flattened and determinate in an alternate pattern. Thallus composed of coenocytic filaments. Thalli were rarely associated with other macroalgae.

Distribution

Maharashtra (Malvan, Bombay), Gujarat (Dwarka, Okha coast, Saurashtra, Porbandar) Tamil Nadu (Krusadai Island, Palk Bay, Gulf of Mannar, Mandapam, Tuticorin, Cape-Comorin, Tiruchendur), Kerala (Kovalam), Karnataka, Goa, Lakshadweep.

Caulerpa serrulata (Forsskål) J. Agardh

Plants are yellowish green in colour and cartilaginous. Thalli creeps on substratum by stolons, giving out rhizoids below and assimilators above. The thalli are small 2- 4 cm tall flattened and divided dichotomously in "Y" shaped. The margins of the blades possess numerous teeth.

Distribution

Lakshadweep Island, Tamil Nadu (Krusadai Island, Tuticorin, Mandapam), Gujarat (Gulf of Kutch), Andaman & Nicobar Islands.

Caulerpa scalpelliformis

Caulerpa serrulata

Codium dwarkense Børgesen

The word *Codium* is Greek and means "skin of the animal" referring to the soft, spongy texture. Plants are green in colour and 14-18 cm tall. The surface of the thalli are finely dotted. Thallus is attached with a disc-like holdfast from which many proliferations arise with main thallus. Thallus is dichotomously branched which are 2-3 cm thick. There is a 2-4 cm gap between two divisions. Vesicles are of two types, slender and broader. Broader vesicle side walls are parallel.

Distribution

Gujarat (Dwarka, Okha, Saurashtra, Porbandar), Maharashtra.

Codium decorticatum (Woodward) Howe

Plants are dark green in colour, attached with well defined discs and 10-30 cm long. Thallus size is very variable, compressed and in some, more or less terete. The thallus often forms a hemispherical or dome-shaped clump. Branches are cylindrical and flattened at each dichotomous fork. Hairs are abundant, ends of branches somewhat broader upwards. Utricles are of variable sizes. Thallus free floating or hanging and subtended in water. In favorable condition very good thalli are seen during low tides.

Distribution

Goa, Tamil Nadu (Mandapam, Krusadai Island), Gujarat (Saurashtra).

Codium dwarkense

Codium decorticatum

Halimeda gracilis Harvey ex J.Agardh.

Plants are grren in colour but slightly calcified. Thalli mostly large, densely intricate populations forming thick cushion on substratum. Color becomes whitish from green after drying. Basal portion not seen branching and branches arise from segment. Segments are flattened up to 5 cm broad, subterete below, cuneate to subreniform in shape some cylindrical segments distributed in different branches, the upper margin of the segments slightly undulate or entire generally trilobed.

Distribution

Andaman and Nicobar Islands, Lakshadweep Island, Tamil Nadu (Krusadai Island, Mandapam, Pamban), Andhra Pradesh (Visakhapatnam).

Halimeda macroloba Decaisne

Thalli are whitish green, solitary, erect and calcified moderately. They are 5-12 cm in height excluding the holdfast, growing on middle and lower intertidal coral reefs. They are attached to substratum by bulbous holdfast, branching mainly di- to trichotomous often become polychotomous in the basal part. Basal segments are commonly compressed trapezoidal in shape, new segments are added apically with a fan-shaped unit or subreniform and the upper margin entire. It was observed that thalli grow in association with other macroalgae like *Enteromorpha* sp., *Hypnea* sp., *Grateloupia filicina* and some epiphytic alga like *Polysiphonia* sp. , (red encrusting) and several minute filamentous algae (red and green) grew on the flat discoid segments of this seaweed.

Distribution

Tamil Nadu (Mandapam, Krusadai Island, Tuticorin, Pamban), Gujarat (Gulf of Kutch), Andaman and Nicobar Islands, Lakshadweep Island.

Halimeda gracilis

Halimeda macroloba

Halimeda opuntia (Linnaeus) Lamouroux

Plants form very dense, stiff-brittle, hemispherical clumps and 10-25 mm in diameter growing in many directions. Plants are usually green but become chalky white when die. Fan structure of frond made up of number of flat, cordate or reniform, strongly calcified segments. These segments are sessile or stalked, cordate, discoid, trilobed or broad in middle and narrowed in both ends, ribbed. Ribs distinct or indistinct.

Distribution

Tamil Nadu (Krusadai Island, Tuticorin, Pamban, Mandapam), Gujarat (Gulf of Kutch), Andaman & Nicobar Islands, Lakshadweep Island.

Halimeda tuna (Ellis & Solander) Lamouroux

Thalli are whitish green and 10-20 cm high. Branched but only in one plane, calcified (but jointed by uncalcified joints) and arise from a short basal stalk attached to the creeping system. Most probably this also plays an important role in the formation of coral reefs. Roots are somewhat bulbous. Lower segments are thicker than upper segments (35-22 mm) may be flat or discoid to elongate, not ribbed with entire margin. Peripheral utricles hexagonal in surface view. Plants are commonly on hardy substrate.

Distribution

Andaman and Nicobar Islands, Lakshadweep, Maharashtra, Tamil Nadu (Krusadai Island, Tuticorin, Gulf of Mannar, Palk Bay), Kerala, Gujarat, (Dwarka, Saurashtra, Okha)

Halimeda opuntia

Halimeda tuna

Udotea indica A. Gepp & E. Gepp

Plants are green in colour with whitish grayish tint on the surface. Thallus either fan-like or funnel shaped with rounded flabellated blades. Attached with a root mass in a small tuft anchored to the substratum. Thallus is erect, 12 cm long, slightly calcified. Stipe simple, 1-3 cm long, terete and , 1-2 mm thick. The flabellum is 6 cm or more in breadth. Flabellum is fan shaped in outlines in young plants with a cuneate base. In older plants flabellum longitudinally folded up with numerous proliferations from the edge of flabellum, folded with deep incisions rotundo-flabellate, orbicular, pandriform obovate. Margins are entire and lobed. Surface of thallus is smooth, transversely zonate with uneven surface. The root mass is well buried in the interstices of the reef-beds formed of corals where there is accumulation of sand and silt. Also found in very shallow lagoons. Alga exposed during low tides, remains unaffected by sunlight.

Distribution

Maharashtra, Tamil Nadu (Tuticorin), Gujarat (Okha, Dwarka, Saurashtra)

Udotea indica

DASYCLADALES, DASYCLADACEAE

Neomeris annulata Dickie

The thallus is cylindrical, sometime club shaped, and small in size up to 3 cm tall. Plants get highly calcified and calcification is more intense from apex downwards, primary branches produced in successive verticils, which number about a hundred. The apical portion of the thallus is tipped with septate deciduous hairs. They are found on rocks and death coral.

Distribution

Tamil Nadu (Krusadai Island, Palk Bay, Tuticorin, Gulf of Mannar, Mandapam), Gujarat (Saurashtra), Lakshadweep Island.

DASYCLADALES, POLYPHYSACEAE

Acetabularia calyculus Lamouroux

Commonly known as "Mermaid's wine glass". This unicellular macroalga is a researchers' favourite. It looks much like a traditional "fan". Plants are clustered, reaching a height of 3-5 cm, stipe rigid and calcified, attached to coral stones and other substrata by a ramified, lobed, rhizomatous base, stipe towards upper end with whorl of scars of deciduous hair branches; on uppermost part, a shallow basin-shaped disc, 4-10 mm in diameter, with upward curving ends with 30-36 gametangia forming disc; apical end of gametangia with 50-80 spherical cysts, up to 95-110 μm in diameter; superior corona cylindrical, somewhat irregular, bearing 2-3 scars.

Distribution

Gujarat (Okha, Veraval), Tamil Nadu (Pamban, Mandapam, Krusadai Island, Lake Pulicat), Maharashtra, Andaman & Nicobar Islands.

Neomeris annulata

Acetabularia calyculus

Phaeophyceae (Brown Algae)

Members of the Phaeophyceae popularly known as brown algae (derived from the Greek word *phaios*, meaning "brown") are almost exclusively marine and attain their greatest abundance, size and diversity in cold temperate water. There are 1,500 species of brown algae which are exclusively found in marine habitats. The brown colour is due to the presence of a brown pigment called fucoxanthin. The largest photosynthetic organism on earth which can grow upto a height of 70 metres tall belongs to this group. The richest area of brown seaweed with its accompanying abundant animal life is the kelp forests which are found in temperate water. In India, we do not have large kelps such as *Laminaria* and *Macrocystis*. Some of the most common brown seaweeds of Indian coasts are *Sargassum*, *Tubinaria, Padina, Dictyota* etc. Some brown seaweeds are used as food and also execellent source of Iodine, Alginic acid, mannitol and laminarin which have several industrial uses.

PHAEOPHYCEAE

Ectocarpus simpliciusculus C.Agardh

Ectocarpus is the prevalent genus and is composed of freely branched, uniseriate filaments differentiated into prostrate and erect system. The prostrate parts are rhizoid like and penetrate to the substrate. Thalli yellowish brown, filamentous, alternately branched, the bases closely entangled, free and feathery above, usually 5-10 cm or more long, growing on lower intertidal rocks or other seaweeds. The branches gradually attenuated sometimes ending in hairs. Plurilocular sporangia are sessile or with short stalk fusiform, gradually attenuating towards the blunt apex, never ending in hairs.

Distribution

Lakshadweep Island (Minicoy), Gujarat, Maharashtra, different parts of Indian Coast.

Ectocarpus simpliciusculus

Dictyopteris australis (Sonder) Askenasy

Thallus is brown in colour with dark spots, erect, with tuberous basal holdfast. The size varies from 10- 40 cm and the blades are flat with ruffled margins may be dentate or sinuate. Stipe covered with short hairs and irregularly dichotomously branched. Midrib is vigorous with irregularly bent veins at both sides of lamina. Hair groups are found on both sides of thallus. Thallus two celled thick but veins are three celled thick. Tetrasporangia found between groups of hair, bright yellow in color and old parts become brown. The older parts of the thallus sometimes get feebly infested with *Melobesia* and other crustaceous red alga.

Distribution

Lakshadweep Island, Karnataka, Maharashtra, Tamil Nadu (Tuticorin), Gujarat (Dwarka, Gulf of Kutch, Saurashtra, Okha), Goa.

Dictyopteris woodwardia (R.Brown ex Turner) C.Agardh

Thallus is pale brown in colour, erect with defined holdfast. Plants erect with well defined holdfast. Fronds 10-15 cm or more in height, flat, 5-7 mm broad, linear, branched repeatedly dichotomously. Midrib distinct extends throughout the length of the frond. Surface of frond perforated on both sides by minute pores which are scattered. Apex of frond bifid with obtuse segments. Margin of frond, dentate, undulating. At base of the frond, the membranous portion worn out by wave action, the midrib alone remains naked. Hairs in groups, and in young parts in a row on both sides of the midrib.

Distribution

Maharashtra (Malwan), Tamil Nadu (Krusadai Island), Gujarat (Gulf of Kutch), West Coast.

Dictyopteris australis

Dictyopteris woodwardia

Dictyota bartayresiana Lamouroux

Plants are erect variable, bushy, densely intermingled masses. Thallus branched, when young but later becomes bent and twisted slightly, dichotomous, equal angled. Internodes 1-4 times as long as broad; margin entire; apex acute, somewhat broadly rounded. Thallus is three celled thick. Tetrasporangia on both sides, either solitary or a few together, scattered over the whole surface. Gametangia on both sides, either solitary or a few together, scattered over the whole surface.

Distribution

Gujarat (Dwarka, Saurashtra, Okha), Maharashtra (Bombay), Tamil Nadu (Pamban, Krusadai Island, Palk Bay, Tuticorin), Goa, Karnataka, Andaman, Nicobar and Lakshadweep Island.

Dictyota dichotoma (Hudson) Lamouroux

Plants are yellowish brown in colour, 10-30 cm long, dichotomously branched forming larger or smaller tufts and entangled masses. Branches more profuse towards upper reaches of fronds and narrow extremities. Tips of branches acute with entire margin. Fructiferous structures, on the middle of the thallus leaving a narrow sterile margin on both sides of it.

Distribution

Tamil Nadu (Krusadai Island, Tuticorin, Pamban, Tiruchendur, Gulf of Mannar, Palk Bay, Mandapam, Pamban), Gujarat (Okha, Dwarka, Saurashtra), Andhra Pradesh (Visakhapatnam), Maharashtra, Goa, Karnataka, Andaman and Nicobar Islands, Lakshadweep Island.

Dictyota bartayresiana

Dictyota dichotoma

Dictyota pinnatifida Kützing

Plants are brown in older parts and light brown at extremities, 10-30 cm long dichotomously branched forming larger or smaller tufts and entangled masses. Fronds, very narrow, much branched, twisted and quite entangled becoming narrower towards extremities. Branches more profuse towards upper reaches of fronds and narrow extremities.

Distribution

Maharashtra (Bombay), West Coast of India, different parts of Indian Coast.

Lobophora variegata (Lamouroux) Womersley ex Oliveira

Plants are commonly known as "Encrusting fan leaf algae". Thalli are orange to dark brown in colour with radial markings, decumbent on rock, ranging from prostrate, resupinate to erect, irregularly lobed blades. The lobes are upto 2.5 cm wide, somewhat clef, becoming broadly fan shaped to suborbicular to kidney shape. Prostrate thalli attached by moniliform rhizoids arising from the ventral surface; margins not inrolled. They are small, thick, prostrate plants as undergrowth of larger seaweeds in the low intertidal; large, erect plants epilithic on coral debris in lagoons; small, thin, prostrate plants on deep subtidal rocks.

Distribution

Maharashtra (Bombay, Malwan), Gujarat (Dwarka, Gulf of Kutch, Okha, Saurashtra), Tamil Nadu (Tuticorin, Krusadai Island, Pamban, Gulf of Mannar, Palk Bay, Mandapam), Andhra Pradesh (Visakhapatnam), Goa, Andaman and Nicobar Islands, Lakshadweep Island.

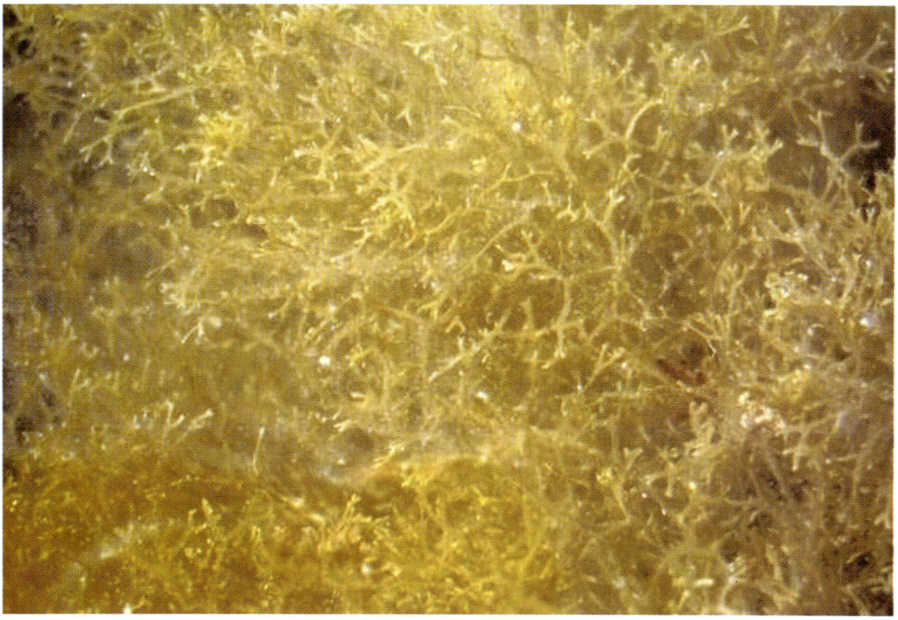

Dictyota pinnatifida

Lobophora variegata

Padina boergesenii Allender & Kraft

Plants are commonly known as "Peacock algae". Thalli are erect, funnel-shaped, with fan-shaped blades which grow on hard substrates like rocks in shallow water. Fronds lightly calcified, hair and sporangial lines with concentric pale and dark transverse bands; dark bands formed by rows of hairs on the ventral as well as the dorsal blade surface, and a distinctive inrolled distal margin basal parts often with rufous patches of rhizoids. Blades distromatic except at base, or blades polystromatic.

Distribution

Kerala, Tamil Nadu (Mandapam, Tiruchendur).

Padina boergesenii

Padina tetrastromatica Hauck

Plants tuft with perennial prostate richly branched basal portion forming a distinct holdfast and attached to substratum by means of tufted rhizoids on dead corals and also on rocky surface. Fronds erect, 12-15 cm height, brownish to olive green in colour, broadly fan-shaped at upper portions, blades frequently lobed and spilt into numerous segments of 1-2 cm broad. Margin is revolute and curved. Zonation caused by rows of hairs and fructification organs in concentric zones. Hairs present in younger thallus, in older ones, either rudimentary or absent. Fronds are thick and leathery in nature.

Distribution

Andaman and Nicobar Islands, Lakshadweep Island, Maharashtra (Bombay, Malwan, Ratnagiri), Tamil Nadu (Gulf of Mannar, Mahabalipuram, Cape Comorin, Palk Bay, Krusadai Island, Tuticorin), Andhra Pradesh (Visakhapatnam), Gujarat (Gulf of kutch, Dwarka, Jalleshwar, Veraval, Saurashtra, Okha), Goa, Kerala (Kovalam), Karnataka.

Padina tetrastromatica

Spatoglossum asperum J. Agardh

Plants are dark brown to dirty green in colour with indistinct holdfast. Thallus is 20-35 cm or more in height. The branches are strap like, dichotomously divided with large and small lobes. The lobes are elongate, linear, lanceolate, attenuated towards base. Apex is acute or rounded, margin sinuate, irregularly dentate with larger or smaller proliferation. Surface smooth and its cells are arranged in more or less distinct rows. Thallus has many epiphytes among which *Melobesia* is common. This seaweed is found on rocks in the subtidal zone of shores exposed to strong wave action.

Distribution

Andaman and Nicobar Islands, Maharashtra (Bombay, Malwan), Tamil Nadu (Krusadai Island, Tuticorin), Gujarat (Gulf of Kutch, Saurashtra, Jalleshwar, Veraval, Okha), Goa, Kerala (Kovalam), Karnataka.

Spatoglossum asperum

DICTYOTALES, DICTYOTACEAE

Stoechospermum marginatum (C. Agardh) Kützing

Plants are large tufts, 12-15 cm or more in height with numerous rhizoids from the base, attaching the plants to the substratum.Thallus erect, with repeated dichotomously branched, without midrib. Thalli are brown in upper parts, tending to be dark brown in the older and basal parts. They are found in rocky pools in lower part of the littoral zone, open sea and also cast ashore.

Distribution

Tamil Nadu (Madras, Tuticorin, Krusadai Island, Mandapam, Killakarai, Tiruchendur, Rameswaram), Kerala, Maharashtra (Bombay, Malwan), Gujarat (Saurashtra, Gulf of Kutch, Okha), Goa, Karnataka.

CHORDARIALES, CHORDARIACEAE

Levringia boergesenii Kylin

Thallus dark to blackish brown in colour, 18-28 cm long, plants are elastic, gelatinous, irregularly branched, grows as a curved hanging chain of flowers or ribbon-like etc. attached to the substratum by tiny basal discs. Lower portion of the axis bears the branches, upper portion has inadequate branches; apical regions have long assimilatory hairs.

Distribution

Gujarat (Dwarka, Okha, Saurashtra), Tamil Nadu, South-West Coast, different parts of Indian Coast.

Stoechospermum marginatum

Levringia boergesenii

Colpomenia sinuosa (Mertens ex Roth) Derbés & Solier

Thallus hemispherical triangular in shape, fragile, sticky appearance, hollow, golden brown plants and crispy in texture. It grows on dead coral in the intertidal and in coral reefs. Young spherical plant becomes irregular, lobed, sinous hollow vesicles which are highly infolded when fully grown in older stage. It may be in solitary or conjugate stage.

Distribution

Lakshadweep Island, Andaman and Nicobar Islands, Karnataka, Maharashtra (Bombay, Ratnagiri), Tamil Nadu (Mandapam, Krusadai Island, Tuticorin), Gujarat (Gulf of Kutch, Saurashtra, Jalleshwar, Veraval, Dwarka, Porbandar, Okha), Goa.

Hydroclathrus clathratus (C. Agardh) Howe

Thallus is yellow to olive brown masses with distinctive perforation, size and shape is irregular net-like in older stage. Spreading thallus loosely attached by the undersurface rocks or coralline pieces and clustered, covering large areas. Sometime interspaced tissue becomes very delicate. Thallus sometimes becomes very fragile. In favorable condition they can grow on neighboring plant as well. Thallus may be thick or thin.

Distribution

Lakshadweep Island, Andaman and Nicobar Islands, Karnataka, Maharashtra (Bombay, Malwan, Ratnagiri), Tamil Nadu (Mandapam, Krusadai Island, Tuticorin), Gujarat (Gulf of Kutch, Porbandar, Jalleshwar, Veraval, Saurashtra, Okha), Goa.

Colpomenia sinuosa

Hydroclathrus clathratus

Iyengaria stellata (Børgesen) Børgesen

Plants are bright yellowish to dark brown in colour, cartilaginous cushions. Plants are smooth, globular, attached by rhizoid. Surface irregularly furrowed, covered with branched knobbly protrusions with irregularly lobed with solid, stellate projections tapering upward and ending in a broad obtuse apex. Thallus solid initially, becoming hollow. Hairs present in scattered bundles.

Distribution

Gujarat (Dwarka, Gulf of Kutch, Okha, Veraval, Saurashtra), Tamil Nadu (Krusadai Island).

Iyengaria stellata

Cystoseira indica (Thivy and Doshi) Mairh

Plant is a dioecious and perennial taxon. The plant is brown in colour, bushy and shows differentiation into holdfast and main axis which branches extensively. The thalli are firmly anchored to the substratum by a cup-like disc. Round stipe represents the main axis which remains hidden due to excessive branching. The air-bladder is prominent and usually occurs on short laterals. The vegetative cycle is characterized by branch shedding, stunted growth, stumpy shoots, dwarf and diminutive holdfast. It grows luxuriantly in the rock pools of the intertidal regions and remains submerged even at low tide.

Distribution

Gujarat (Okha, Saurashtra, Veraval), Maharashtra.

Cystoseira indica

Cystoseira myrica (S.Gmelin) C. Agardh

The bushy, perennial thallus (30-50 cm high) consists of several main axes arising from an irregularly expanded disc. Axes are ramified in alternate distichous pattern. It is a perennial alga, growing singly in semi-exposed and sheltered habitats in the sublittoral zone, to at least 15 m depth. The branches of the perennial thallus possess basal spiniferous, which are totally covered by epibionts (plants and animals), small shells and grains of sand.

Distribution

Gujarat (Okha), West Coast.

Hormophysa cuneiformis (J. Gmelin) P. Silva

Plants erect, with a conical well formed holdfast, light brown in colour. The axis develop three sided much branched irregularly, serrated winged blades. Lateral branches arise from the rachis, few elliptical vesicles are embedded within the former rachis. Conceptacles are abundant on the narrower wings of the ultimate and subultimate branches. Cryptostomata ovate scattered on both surfaces of foliar wings. Colour brown, turning to paler red on exposure, blackish in dry state. Substance coriaceous to cartilaginous, thin.

Distribution

Tamil Nadu (Pamban, Krusadai Island, Tuticorin, Mahabalipuram, Gulf of Mannar, Palk Bay, Mandapam), Gujarat (Okha, Saurashtra), West Coast of India, Andaman and Nicobar Islands, Indian Coast.

Cystoseira myrica

Hormophysa cuneiformis

Sargassum cristaefolium C. Agardh

Plant is dark brown in colour, attached to substratum with a strong disc shape holdfast. Frond is erect, main axis is smooth but terete at the base. Branches alternate; leaves 1-2.5 cm long obovate and obliquely roundish at the lower margin asymmetrical at the base, with dentate margin. Vesicles sparsely present, spherical and with pointed pedicels. Receptacle loosely aggregated with serrated margins.

Distribution

Lakshadweep Island (Minicoy), Maharashtra, (Vengurla), Andaman and Nicobar Islands (Port Blair), Tamil Nadu, Indian Coast.

Sargassum cristaefolium

Sargassum cinereum (J. Agardh)

Plants are short thallus with stout main axis, primary branches are terete, smooth and bisect with secondary branches and branchlets. Basal leaves are oblong, membranaceous about 2.5 - 3 cm long, 7 - 8 mm broad, rounded at the apices with dentate margins. Branchlet leaves lanceolate and cuneate at the base. Vesicles are spherical and obovate in shape.

Distribution

Gujarat (Gulf of Kutch), Maharashtra, Goa, Karnataka, Kerala, Lakshadweep Island, West Coast.

Sargassum cinereum

Sargassum johnstonii Setchell & Gardner

Plants are rough textured with holdfast arise from the main axis that bears the long primary branches, which in turn possess many secondary branches. This secondary branches bear spirally arranged leaves. The leaves are sessile, denticulate margins, distinct midrib and narrow symmetrical base with obtuse apex. The leaf bears cryptoblasts whose paraphyses are multicellular and protrude through the ostiole. The thalli grow luxuriantly and vigorously on rocks, boulder and hard substrata as well as deeper or shallower lagoons which are constantly agitated by current.

Distribution

Gujarat (Okha, Jalleshwar, Veraval, Saurashtra), Maharashtra, Tamil Nadu (Gulf of Mannar), Bengal.

Sargassum johnstonii

Sargassum linearifolium (Turner) C. Agardh

Plants dioecious with broadly lanceaolate leaves, linear leaves up to 10 cm long with acute apex to obtuse which are forked sometimes. Cryptostomata scattered on the surface and margin of the leaves. Female receptacle racemose to cymose, with short stalk while male receptacle racemose to cymose, compressed and winged and spines at margin and apices.

Distribution

Gujarat, Indian Coast.

Sargassum swartzii C. Agardh

Thalli are dark brown, arising from discoid holdfasts and growing on subtidal rocks. Main axis cylindrical, very short and giving rise to several primary erect flattened branches. Secondary branches with numerous branchlets, bearing leaves, vesicles and receptacles. Lower leaves slightly lanceolate, with sparse teeth at the margins. Upper leaves of the branches elongated-linear, midrib indistinct, toothed or entire at the margins. Cryptostomata rare, arranged along the margins. Leaves accompanying the receptacles very slender and linear. In mature thalli only upper linear or cylindrical leaves present. Vesicles ovate or elliptical, very variable in size, often crowded at the apex.

Distribution

Gujarat (Okha, Saurashtra), Tamil Nadu (Cape Comorin, Madras, Krusadai Island), Maharashtra, Andaman and Nicobar Islands, South India West Coast, Goa.

Sargassum linearifolium

Sargassum swartzii

Sargassum tenerrimum J. Agardh

Plants are yellowish in colour, pyramidal in shape with basal adhesive disc or holdfast from which arises a short primary axis bearing several branches together. The axis is rounded and glabrous, secondary branches fork repeatedly in their turn. Leaves are wide and tapering at the end. They are thin, translucent and linear-lanceolate, with toothed margin the midrib is indistinct. Cryptostomata are freely scattered on the leaves and receptacles while they are rare on the axis. The receptacles are freely-branched structures.

Distribution

Maharashtra (Bombay), Gujarat (Okha, Adatra reef, Cannanore, Dwarka, Saurashtra, Jalleshwar, Veraval, Gulf of Kutch), Tamil Nadu (Cape Comorin, Krusadai Island), Andhra Pradesh (Waltair, Visakhapatnam), Karnataka, Kerala (Travancore), Andaman Island, East Coast, West Coast, Goa, Indian Coast.

Sargassum tenerrimum

Sargassum wightii Greville

Plants are yellowish brown which grow up to 15-25 cm in height. Upper portion highly branched, axes cylindrical with well defined holdfast, leaves 5-8 cm long and 2-9 mm broad with tapering end. Plants bear large spherical vesicles. Receptacles in clusters and repeatedly branched.

Distribution

Tamil Nadu (Cape Comorin, Krusadai Island, Tuticorin, Gulf of Mannar, Madras, Palk Bay, Mandapam, Pamban, Muthupet Estuary, Killakarai, Rameswaram), Karnataka (Mangalore), Goa, Andhra Pradesh, Maharashtra, Andaman and Nicobar Islands Kerala (Travancore, Kovalam), West Coast, South India, South West Coast).

Sargassum wightii

Turbinaria conoides (J. Agardh) Kützing

Turbinaria conoides is a compact, cone-like and radially organized plant. The thalli are light brown, main axis and branches slightly compressed. Branches several, branchlets turbinate, triangular and dentate. Vesicles inflated in the middle of the blade. The thalli are firmly anchored to substratum primarily by a disc which is later supplemented by a number of terete spreading branches. This species is attached to rocks in reefs by a few branched rhizoids.

Distribution

Andaman and Nicobar Islands, Lakshadweep Island, Maharashtra, Kerala, Tamil Nadu (Krusadai Island, Tuticorin, Mandapam, Pamban, Gulf of Mannar), Gujarat (Okha), Andhra Pradesh (Visakhapatnam).

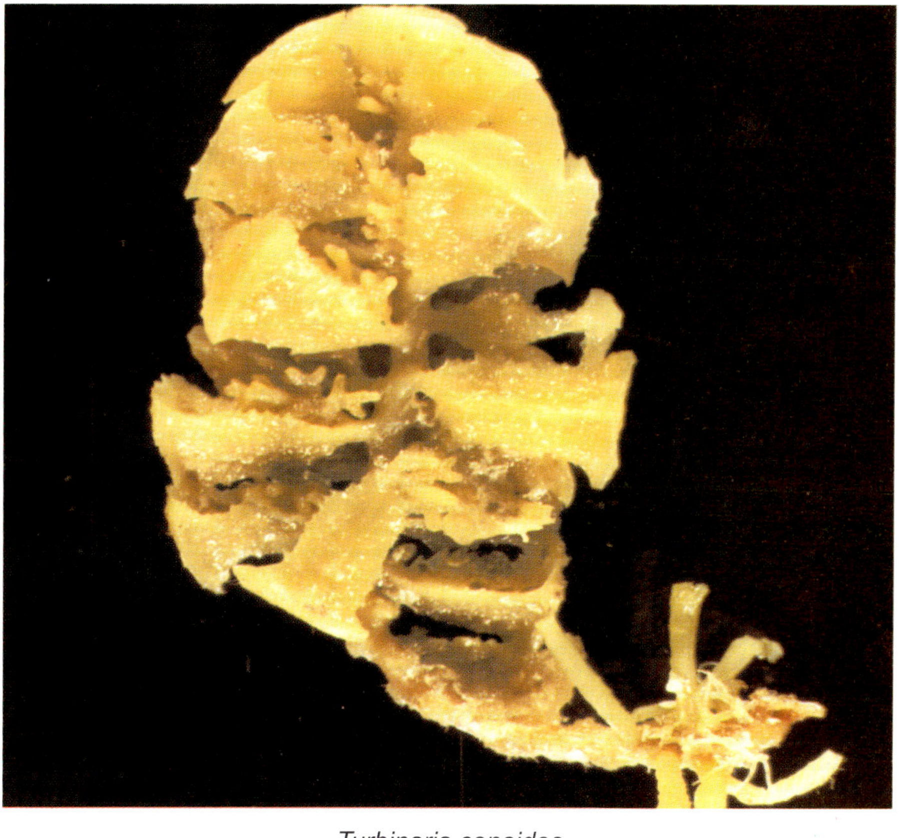

Turbinaria conoides

Turbinaria decurrens Bory de Saint Vincent

Plants about 15-20 cm high, rarely branched, leaves fleshy, pyramidal, long, stalk short and terete. The upper portion longitudinally ridged and the distal end inflated and triangular and the margins provided with numerous minute spines; large vesicles. Receptacles compact, racemose and about half of the length of laterals.

Distribution

Andaman and Nicobar Islands, Lakshadweep Island (Minicoy), Tamil Nadu (Mandapam), Andhra Pradesh (Visakhapatnam).

Turbinaria ornata (Turner) J. Agardh

Thalli are yellowish brown, with numerous branches in all directions, attached by discoid holdfasts to subtidal reefs of rocks. Main axis is erect and cylindrical, usually with several branches. The leaves are coarse and firm, longitudinally ridges obtuse, marginal blades conspicuous, rounded triangular, with short teeth along the margins. Receptacles racemose, located about one-third the distance from the base of the stalks. The plants were found growing entangled by *Ulva reticulata* and other macroalgae at its vicinity were *Sargassum* sp., *Gelidiella* sp., *Portiera* sp., *Hypnea* sp. etc.

Distribution

Gujarat (Dwarka, Okha, Saurashtra), Tamil Nadu (Gulf of Mannar, Palk Bay, Mandapam, Tuticorin), Andhra Pradesh (Visakhapatnam), Andaman and Nicobar Islands, Lakshadweep.

Turbinaria decurrens

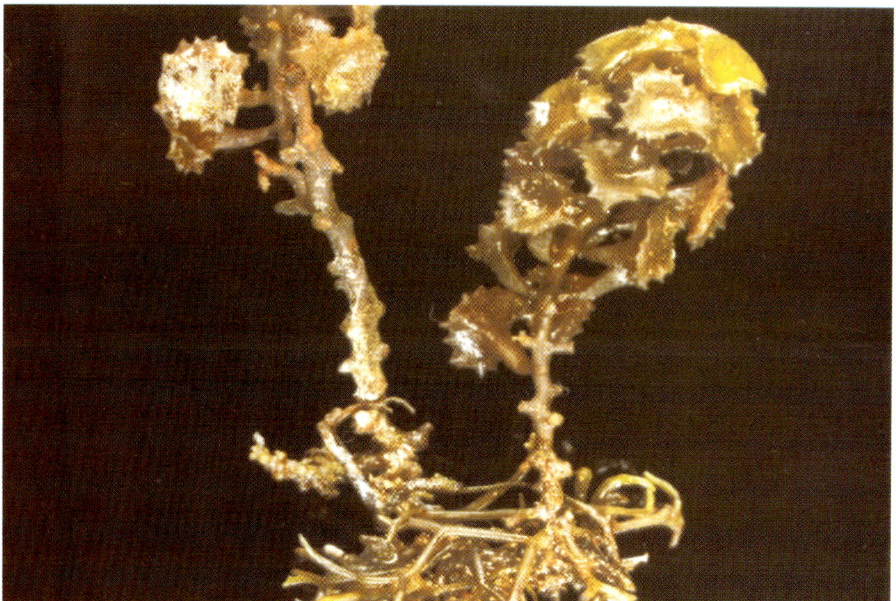

Turbinaria ornata

Turbinaria triquetra (J. Agardh) Kützing

The thallus is highly seasonal, tropical environment. The species are large, erect intertidal seaweeds, branched or unbranched, firmly attached to the substratum by means of spreading rhizomatous structures. The main axis is covered with closely arranged laterals. The lateral foliar structure is provided with basal terete stalk and gradually expanded upper portion with pyramidal, turbinate, or peltate appearance. Thalli consist of a holdfast system formed by branched haptera, from which primary axes arise.

Distribution

Andaman and Nicobar Islands.

Turbinaria turbinata (Linn.) Kuntze

Thallus tough, leathery, erect, cylindrically shaped, dark brown. Blades clustered, apex flat or convex, pyramid-shaped. Transverse section through blade triangular, margins toothed. Cryptostomata numerous, scattered, uniporate. Found growing on coral rubble.

Distribution

Gujarat(Okha), Lakshadweep Island, Andaman Island.

Turbinaria triquetra

Turbinaria turbinata

Rhodophyceae (Red Algae)

Rhodophyceae (derived from the Greek word ***rhodon*** "red rose") is by far the largest and most diversified group amongst algae. More than 4,000 species, have been reported mostly from tropical marine habitats. Red algae range in morphology from simple unicells to unbranched and branched filaments to complex multiaxial uprights and crusts. Rhodophytes occupy the entire range of depths inhabitable by photosynthetic plants, from high-intertidal regions upto several depths. The plants look red because of presence of the dominant pigments r-phycocynin and r-phycoerythrin. Some of the most commercially important seaweeds belong to this group. Most common red seaweeds are *Porphyra, Gracilaria, Grateloupia, Kappaphycus*, *Laurencia*, *Polysiphonia* etc. *Porphyra* popularly known as Laver in England, Nori in Japan, Kim in Korea, Zicai in China is used as food and globally a US$ 1.8 billion industry. *Gracilaria* and *Gelidium* are very good source of agar. *Kappaphycus* and *Grateloupia* are good source of carrageenans.

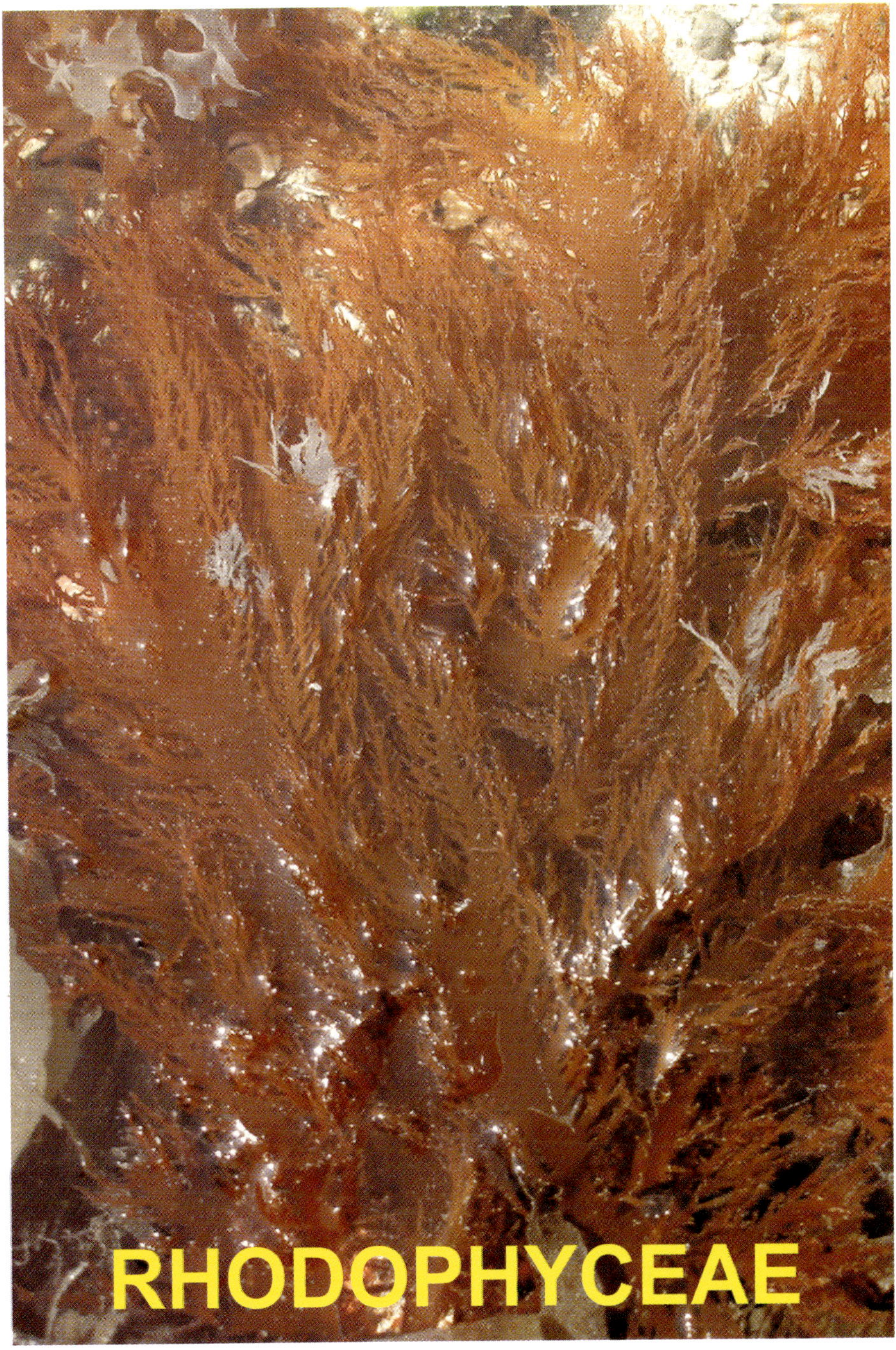
RHODOPHYCEAE

BANGIALES, BANGIACEAE

Extensive growth of *Porphyra sp.* on rocky coast of Goa and Maharashtra. *Porphyra sp.* is a temperate genus but few tropical species grow in India. *Porphyra sp.* in India mostly grow for a short period during monsoon season from July-September.

Porphyra sp. at Dona Paula Goa

Extensive growth of *Porphyra sp.* at Shreewardhan Maharashtra

Porphyra vietnamensis T.Tanaka and Pham-Hoáng Hô

Blades are membranous, lanceolate to linear-lanceolate, purplish-pink in colour. Plant exhibits great deal of plasticity in its morphology. The blades are mostly ribbon like but sometimes have broad robust thalli, a large number of bladelets arise from common holdfast. Sometimes the blade is highly unorganized having several bladelets which have folded margins. The thalli are usually 3-15 cm in height but sometimes grow upto 40 cm whereas they are 1.5-3 cm broad. The thalli are attached to the substratum by discoid holdfast which has large number of rhizoids. The thalli of *P. vietnamensis* like other *Porphyra* species have been clearly differentiated into three distinct regions such as holdfast, central vegetative and outermost fertile region.

Distribution

Tamil Nadu (Madras, Mandapam), Andhra Pradesh (Visakhapatnam), Karnataka, Maharashtra, Goa.

Porphyra vietnamensis growing alongwith *Ulva lactuca*

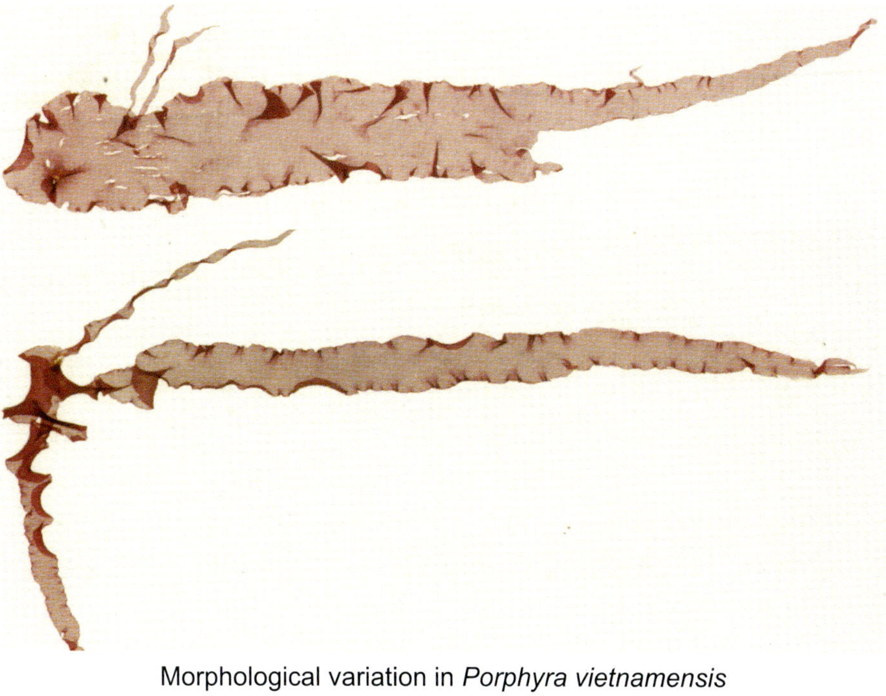

Morphological variation in *Porphyra vietnamensis*

Porphyra kanyakumariensis V. Krishnamurthy and Baluswami

The thallus is broad, membranous, undulate, lanceolate in shape and pale violet in colour. The blades are 15-33 cm long and 2.0-5.6 cm braod and grow luxuriantly in upper intertidal region. The blades are attached to the substratum by discoid holdfast through a large number of rhizoids. The entire blade starting from the holdfast region to the upper fertile part is covered with thick mucilaginous layers. The margins of the blade are serrated with single or multicelled spines which are 6-8 µm long. The holdfast region has large round, oval, elliptical cells which have long clear, thin protuberances oriented towards the base. These cells are embedded in thick intercellular matrix.

Distribution

Kerala (Kovalam), Tamil Nadu (near Kanyakumari).

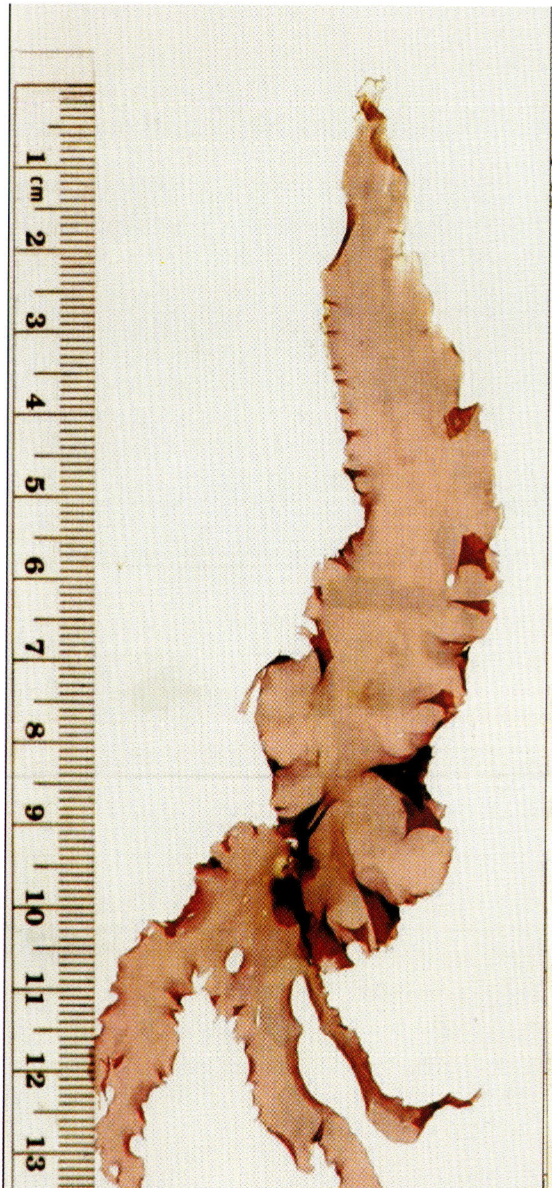

Porphyra vietnamensis

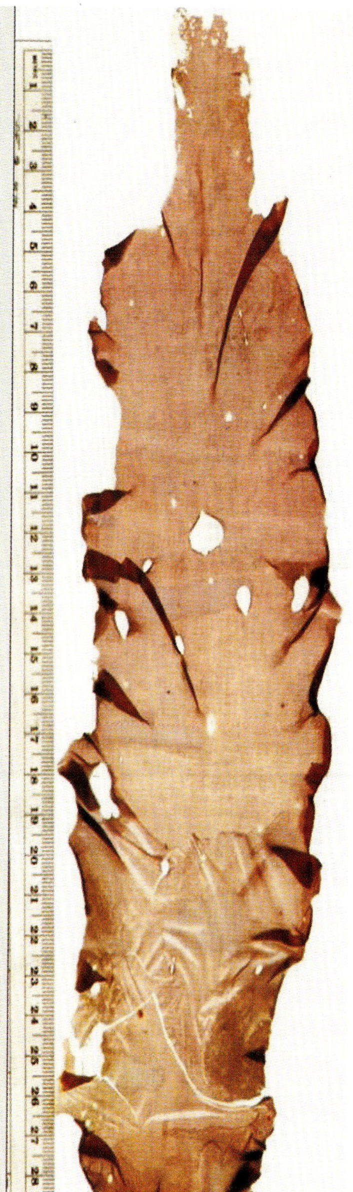

Porphyra kanyakumariensis

Porphyra suborbiculata Kjellman.

The plants grow abundantly in the upper intertidal zone on the granite rocks and as an epiphyte on *Enteromorpha intestinalis* and *Ulva lactuca*. The plants are small, 0.6-1.5 cm long and 0.5-2 cm broad and deep pinkish blue in color. The thalli are round, reniform to orbiculate in shape with undulate margins. The margins of the blade are serrated with distinct single celled spines. The spines have swollen rounded tips. The spines are 3-9 μm in length. The blades are attached to the substratum by discoid holdfast. The holdfast region is covered with thick mucilaginous secretion. The cells in the holdfast region are small, oblong to capitate in shape. The cells of the holdfast region do not have the typical pear-shaped heads with long thin tails as seen in other three *Porphyra* species.

Distribution

Kerala (Quilon Coast).

Porphyra chauhanii Anil Kumar & Panikkar

The blades are membranous, cordiform and broadly expanded with serrated margins. The plants are 7-13 cm in length and 5-10 cm in breadth and reddish brown in colour. The thalli grow abundantly on the rocks in the upper tidal region in close association with *Enteromorpha intestinalis* and *Ulva lactuca.* The blades are attached to the substratum with disc-shaped holdfast. The thalli-like other *Porphyra* species have distinct holdfast, central vegetative and fertile regions. The margins of the blade in this region have spinulate processes. The thallus is monoecious and male gametangia can be distinguished as light yellow coloured narrow patches which are intermixed with dark reddish zygotosporangia.

Distribution

Kerala

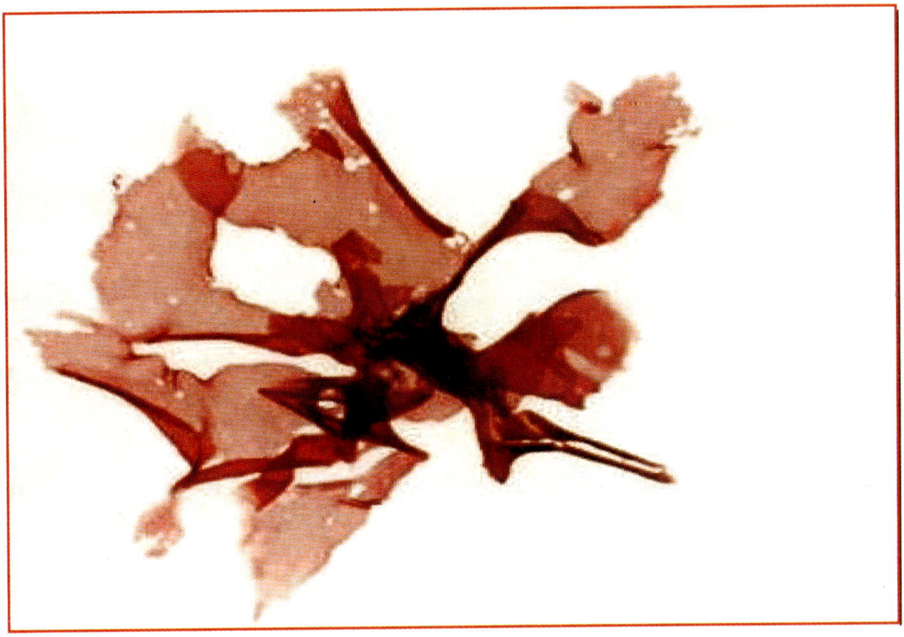

Porphyra suborbiculata

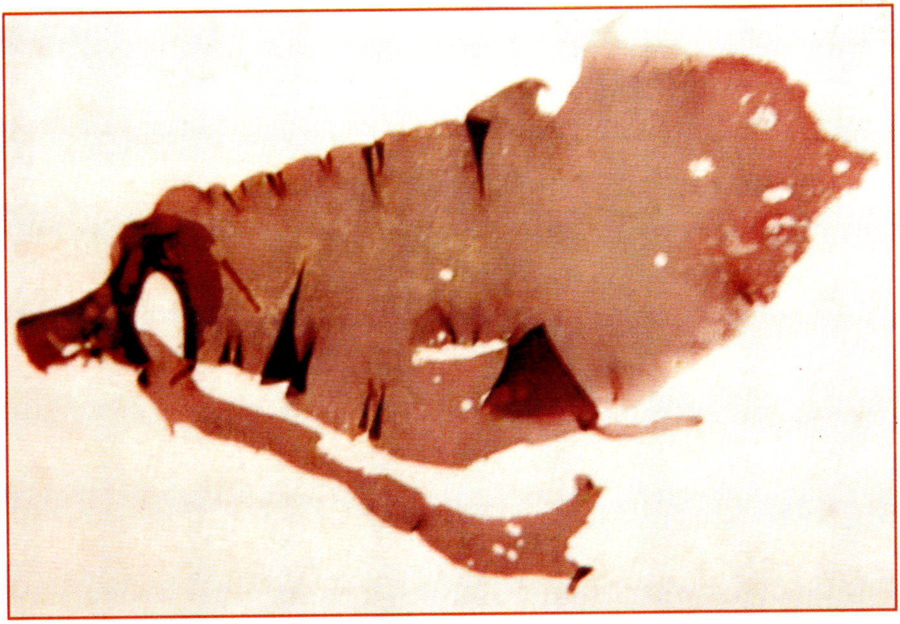

Porphyra chauhanii

Scinaia carnosa (Kützing) J. Agardh

Plants are dark red in colour, upright upto 17 cm high, murky, several times dichotomously branched, thin, 2 mm wide when dry, highly constricted at irregular intervals. Thallus cylindrical with narrow axils, central axils inconspicuous when dried. Utricles even at top.

Distribution

Kerala (Cape Comorin), Tamil Nadu, South-West Coast of India.

Scinaia moniliformis J. Agardh

Thallus brownish red, large sized, turgid, 15 cm or higher, 2-3 mm broad, repeated forked sometimes constructed at the nodes. Dichotomously branched more than 7-8 times and plants show clustered, articulated and beaded appearance. Lower portion is darker than upper. Joints form and size are variable. Mucilage found in the cavity of joints. Peripheral cells are polygonal in surface view, colorless, 40-45 m long and 17-19 m broad. Cystocarp scattered over the surface of the joints.

Distribution

Gujarat (Okha, Saurashtra, Gulf of Kutch), Andaman and Nicobar Islands.

Scinaia carnosa

Scinaia moniliformis

NEMALIALES, LIAGORACEAE

Liagora albicans Lamouroux

Plants erect, gelatinous, cylindrical or subcylindrical, broad at the base and tapering to the tips; irregular dichotomous or pinnate branching; calcification in various degrees, multiaxial and solid, with a central medullary filaments; cortical filaments repeatedly dichotomously forked, cells long, moniliform, shorten and rounded at the advance of periphery, sexual plants dioecious or monoecious.

Distribution

Tamil Nadu (Madras, Mahabalipuram, Mandapam), Gujarat, Andhra Pradesh (Visakhapatnam), Andaman and Nicobar Island, Lakshadweep Islands, Indian Ocean.

AHNFELTIALES, AHNFELTIACEAE

Ahnfeltia plicata (Hudson) Fries

Thallus deep red to purplish black in colour, erect 10 cm tall, repeatedly dichotomously branched, cylindrical in structure, upto 0.5 mm. Thallus shows occasional proliferations, short 1-2 dichotomous branching.

Distribution

Maharashtra (Bombay), Goa, different parts of Indian Coast.

Liagora albicans

Ahnfeltia plicata

Gelidium pusillum (Stackhouse) Le Jolis

Plants small, tough and cartilaginous, red in colour, 6-7 mm high, horizontal stolons as prostrate portion irregularly branched without midrib, attached by rhizoids, the erect fronds simple or rarely branched. Plants are light to dark red, and form turf-like masses on sandy rocks in the subtidal zone of moderately wave-exposed shores.

Distribution

Gujarat (Dwarka, Jalleshwar, Veraval, Okha), Maharashtra (Bombay), Andhra Pradesh (Visakhapatnam), Tamil Nadu (Gulf of Mannar, Mandapam, Tiruchendur), Kerala (Kovalam), Karnataka (Mangalore), Odisha, Goa, Andaman and Nicobar Islands, Lakshadweep Island.

Gelidium pusillum

Gelidiella acerosa (Forsskål) J. Feldmann & G.Hamel

Thalli are purplish, cartilaginous in texture, formed by the several tufts, cylindrical erect axis up to 5-7 cm high with a basal portion. They are attached to the substratum by rhizoids on rocks, dead corals with basal creeping rhizome. Branching usually distichously arranged opposite or subopposite pinnate.

Distribution

Gujarat (Okha, Adatra, Jalleshwar, Veraval, Saurashtra, Dwarka), Tamil Nadu (Krusadai Island, Tuticorin, Gulf of Mannar, Mandapam, Palk Bay, Kilakkarai, Pamban), Kerala (Turichendur), Andhra Pradesh (Visakhapatnam), Lakshadweep Island (Minicoy), Andaman and Nicobar Islands.

Gelidiella acerosa

Gracilaria corticata (J. Agardh) J. Agardh

Gametophytic plants grow together in dense tufts and arise from hard holdfasts that are attached firmly to rocky substrata. Thallus reaches about 10-15 cm in length and is rigid, cartilaginous for major part except for the ramuli tip that repeatedly branches in either bi- or tri- partite manner. The segments width are 1 or 2 mm but occasionally reach upto 4 mm. The thallus is violet red in colour. *Gracilaria corticata* is dioecious plant. Both male and female plants are morphologically dissimilar. The male plants can be readily distinguished by their pale red colour; narrow and linear branches that are repeatedly bi- or tri-partite. The apices have sharp and tapering ends. The spermatangial conceptacles appear as fine dots and are scattered over the middle part of the thallus. A young female plant shows many flattened, dichotomously branched fronds arising from the holdfast that deeply impregnated the calcareous rocks. The apices of the female plants are acute. The carpogonia are born near the apices of the branches and their prolonged trichogynes protrude and even provide asperity to the thallus surface. The apical portion of the thallus has striated surface and also shows wavy crest and tufts.

Distribution

Andaman and Nicobar Islands, Maharashtra (Bombay), Kerala (Kovalam), Tamil Nadu (Tuticorin, Kalpakkam, Mandapam, Tiruchendur, Pudumadam, Gulf of Mannar), Gujarat (Dwarka, Saurashtra, Jalleshwar, Veraval, Okha), Andhra Pradesh (Visakhapatnam, Gangavarum), Goa, Karnataka (Mangalore), South East Coast, West Coast, South West Coast.

Gracilaria corticata

Gracilaria corticata (J. Agardh) J. Agardh var. ***cylindrica*** Umamaheshwara Rao.

The plants are strongly attached to the rocky substrata. Thalli are fleshy, dark red to yellowish red in colour, upto 12 cm in height, attached by a small disc. Main axis thin below becoming thicker above in thickness and cylindrical. Branching dichotomously, alternately or irregular, coming out on all sides with pointed or spinous apices.

Distribution

Tamil Nadu (Mandapam), Andaman & Nicobar Islands.

Gracilaria corticata var. *cylindrica*

Gracilaria crassa Harvey ex J. Agardh

Plants are short, thick, cylindrical, about 5 cm in height, 0.2 cm thick and greenish in colour. It has an irregular to discoid holdfast, which forms a dense cushion-like structure on its substratum. Branching is dichotomous and the branches bend towards the substratum. Thallus is fleshy and segmented near apices. Male plants are identical to the tetrasporic plants and it is difficult to identify through naked eyes. Male plants are slightly thicker and shorter than the other thalli. The cyctocarpic thallus can be easily distinguished due to the presence of dome-shaped cyctocarps. The cystocarp are about 0.2 cm wide and sparsely scattered over the thallus.

Distribution

Lakshadweep Island (Minicoy), Kerala (Cape Comorin), Tamil Nadu (Tuticorin, Mandapam, Krusadai Island), Andaman & Nicobar Islands.

Gracilaria crassa

Gracilaria edulis (S. Gmelin) P. Silva

The thallus is reddish brown in colour, cylindrical, 20-25 cm in height and branching pattern is 3-4 order, sub-dichotomous in nature. They form clump with attenuated base and they are attached to coral stones or to the substratum with discoid holdfast. It was observed that thallus entangled with other macroalgae like *Ulva reticulata, Enteromorpha* sp.

Distribution

Tamil Nadu (Tuticorin, Krusadai Island, Pudumadam, Pamban, Mandapam, Gulf of Mannar, Madras), Maharashtra (Bombay), Odisha (Chilika lake), Kerala, Lakshadweep (Minicoy), Andaman and Nicobar Islands, Andhra Pradesh (Visakhapatnam, Waltair Coast).

Gracilaria edulis

Gracilaria eucheumatoides Harvey

Plants are dark red to deep purplish in colour, up to 5-8 cm length and attached by heptera. The thalli are thick, fleshy, compressed, irregular and dichotomously branched. The cystocarps are protruded and scattered all over the thallus.

Distribution

Andaman and Nicobar Islands (Chidiya Tapu), Indian Coast.

Gracilaria foliifera (Forsskål) Børgesen

The thalli are flat, translucent, dull purple or reddish brown in colour and measure up to 10-15 cm. Plants are dichotomously or polydichotomously branched with alternate branched fronds giving rise to a bushy appearance. Frond is membranous, brittle with compresses stipe about 1-2 cm width near base. Stipe expands gradually into a blade, which is up to 1mm thick and upto 1cm broad. The margins of the blade are frequently proliferous. Cystocarps are elevated, subsperical, constricted at the base about 0.1-0.2 cm width, scattered over the thallus.

Distribution

Tamil Nadu (Tuticorin, Cape Comorin, Krusadai Island, Mandapam, Gulf of Mannar, Pudumadam, Palk Bay, Tiruchendur), Gujarat (Saurashtra, Okha) Maharashtra (Bombay), Kerala, Andhra Pradesh (Visakhapatnam, Waltair Coast) Andaman and Nicobar Islands and Lakshadweep Island.

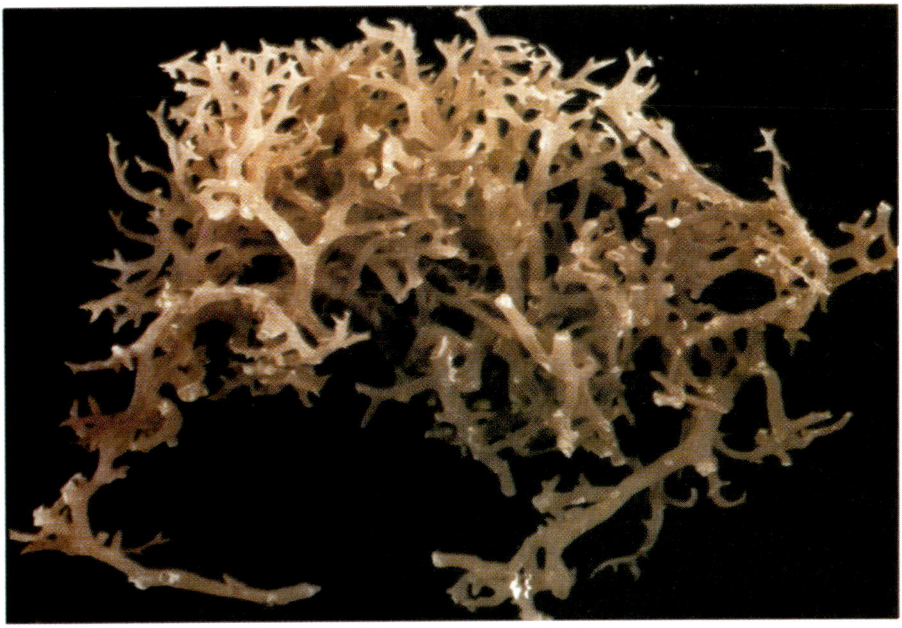

Gracilaria eucheumatoides

Gracilaria foliifera

Gracilaria verrucosa (Hudson) Papefuss.

Gracilaria verrucosa has two types of morphotype *i.e.* floating type and attached type. Mature thalli of floating types are deep brown in colour and smaller in size, measure between 6-12 cm but sometimes can even grow up to 20 cm in length. Branching is sub-dichotomous and attenuated at tip and branches up to 4th order.

Mature thalli of the attached type plants are reddish brown or purple in colour, round and can grow up to 65 cm in length. However, thalli measuring up to 5 ft. in height have been reported from Kalijai Island. The plants are attached to substratum with the help of a small discoid holdfast. Branching is profuse which make the plants look bushy in appearance and brancing is upto 3rd order. Only the attached type is fertile.

Distribution

Andaman and Nicobar Islands, Odisha (Chilika lake), Tamil Nadu (Tuticorin, Porto Novo, Mutupet Estuary, Madras, Krusadi Island, Mandapam, Pudumadam, Gulf of Mannar), Maharashtra (Bombay), Gujarat (Saurashtra), Andhra Pradesh, Goa, Pondichery.

Gracilaria verrucosa

GRACILARIALES, GRACILARIACEAE

Gracilaria is an important source of agar. The pictures show cultivation of *Gracilaria verrucosa* at Chilika Lake, Odisha.

Rope cultivation of *Gracilaria verrucosa* at Chilika Lake, Odisha

Pond culture of *Gracilaria verrucosa* at Chilika Lake, Odisha

Gracilariopsis megaspora Dawson

The thalli are reddish purple in colour and turn pale yellow when exposed to the sun. The thalli are long, rounded and grow upto 50 cm in height. The thallus is attached to the substratum by a small discoid holdfast. The branching is irregular and upto 4th order. Secondary branches are smaller in size as compared to the main axis. Sometimes branches of all order are surrounded with small proliferations.

The gametophytic plants are dioecious. Male plants are comparatively smaller than the female plants. The cystocarpic plants are denser than the tetrasporic plants. The cystocarpic plants can be easily distinguished to the naked eyes as the cystocarps are formed as bulging protuberances on the thallus surface. The mature cystocarps are globose or sub-spherical in shape.

Distribution

Odisha (Chilika lake, Balugaon), Andhra Pradesh.

Gracilariopsis megaspora

Asparagopsis taxiformis (Delile) Trevisan

Plant is red to bluish violet in colour, has a fluffy appearance which looks like a Christmas tree, with a creeping rhizomatous portion bearing descending rhizoidal holdfasts and erect secondary branches; the creeping part usually much branched; erect branches almost naked up to the middle, densely covered with branches and branchlets subopposite pinnate, commonly grows in shallow subtidal habitats with heavy water motion.

Distribution

Gujarat (Dwarka, Saurashtra, Gulf of Kutch, Okha), Tamil Nadu (Mandapam), Maharashtra, Goa, Andhra Pradesh, Lakshadweep Island, different parts of Indian Coast.

Asparagopsis taxiformis

CRYPTONEMIALES, CORYNOMORPHACEAE

Cryptonemia undulata Sonder

Plant erect, 12-15 cm high, light red in colour, firm and elastic in consistency; basal attachment disc small, erect fronds caespitose, stipitate; the stipe rigid, cylindrical, dichotomous, expanding above into cuneate oblong linear blades which are up to three times furcate, the angle between divarications rather obtuse; blades 1-2 cm wide. The margin pronouncedly undulate the lobate; midrib prominent in lower portions, gradually disappearing in the upper portions; apices blunt; proliferations common, variable in number, arising from the midrib, obovate oblong and anervate, but otherwise similar to the main blades.

Distribution

Gujarat(Okha, Dwarka), Different parts of Indian Coast.

CRYPTONEMIALES, HALYMENIACEAE

Grateloupia filicina (Lamouroux) C. Agardh

Mature thalli are deep reddish brown in colour, slimy and grow upto 12 cm. The plants turn light green in colour when exposed to sunlight. The thallus is highly branched with several upright branches arising from common discoid holdfast. The thallus is rounded at the basal region and becomes flattened towards the apical region. Branches appear from the flattened area of the main axis. Branches are long, linear and short in height than the main axis. The branches either end up into a sharp point or bifurcate only once. The axis is beset with numerous proliferations that are irregularly arranged along the margins. The proliferations are terete and slightly constricted at the base but pointed to round at the tip. The gametophytic plants are either monoecious or dioecious. Male plants are smaller; more sparsely branched and bear fewer proliferations than the female plants.

Distribution

Kerala, Odisha (Chilika Lake), Tamil Nadu (Tuticorin, Mahabalipuram, Cape Comorin, Mandapam, Tiruchendur), Maharashtra (Bombay, Malwan), Andhra Pradesh (Visakhapatnam), Goa (Anjuna), Karnataka (Mangalore).

Cryptonemia undulata

Grateloupia filicina

Grateloupia indica Børgesen

Thallus highly lobed and are attached to substratum by a very small basal disc. Stipe short, compressed, cuneate, expanding in broad thallus; thallus flat, oblong to linear, extensively divided, tough, slippery and large; margin irregularly sinuate, tips gradually tapering. Cystocarp dispersed throughout the thallus, subspherical to spherical, occasionally flattened.

Distribution

Gujarat (Okha, Gulf of Kutch), and different parts of Indian Coast.

Grateloupia indica

Grateloupia lithophila Børgesen

Plants of *Grateloupia lithophila* are green to olive brown in colour, grow up to height of 28 cm and exhibit considerable morphological variations. The surface of the thallus is highly lubricous and gives slippery touch. The thallus grows either singly or in a group up to 12 individuals in thick short clump. The thallus gets firmly attached to the substratum through a discoid holdfast. The main axis is round at bottom and measures up to 0.5 cm in height, which gradually flattens towards the apex. The width of main axis varies from 0.5-2 cm. The thallus is simple or divided, linear-lanceolate, tapering from the middle towards both ends.

Distribution

Tamil Nadu (Tuticorin, Krusadai Island, Madras, Mandapam, Tiruchendur), Kerala (Kovalam, Cochin), Maharashtra (Bombay), Karnataka (Mangalore) Andhra Pradesh (Visakhapatnam), Goa.

Grateloupia lithophila

Halymenia dilatata Zanardini

Plants erect in structure with variable height, rose to red in colour and gelatinous in consistency; basal attaching disc small, the single (occasionally two, rarely more) erect frond stipitate; stipe very short, horny and subterete, expanding abruptly into the broad lamina; lamina as long as broad or at times longer, variable in shape, ranging from flabelliform to oblong, entire or palmatifid, the lobes often imbricate, giving the blades a twisted appearance; margin of blades with numerous small denticulate proliferations and frilled in appearance, the surface usually maculate, the patches irregular and soriform; occasionally a few large proliferations, similar to the main blade, also present.

Distribution

Tamil Nadu (Tuticorin, Mahabalipuram), Maharashtra (Malwan), Indian Coast.

Halymenia porphyraeformis Parkinson

Plants upto 48 cm high, rose red in colour, tough and elastic in consistency. Basal disc small and giving rise to single erect stipitate fronds, stipe short, about 1 cm long, expanding abruptly into the suborbicular lamina with a cordate base and sinuate or wavy margins, often becoming lobed in the upper portions.

Distribution

Gujarat (Okha, Dwarka, Okha Port, Saurashtra Coast), Maharashtra, Tamil Nadu and Indian Ocean.

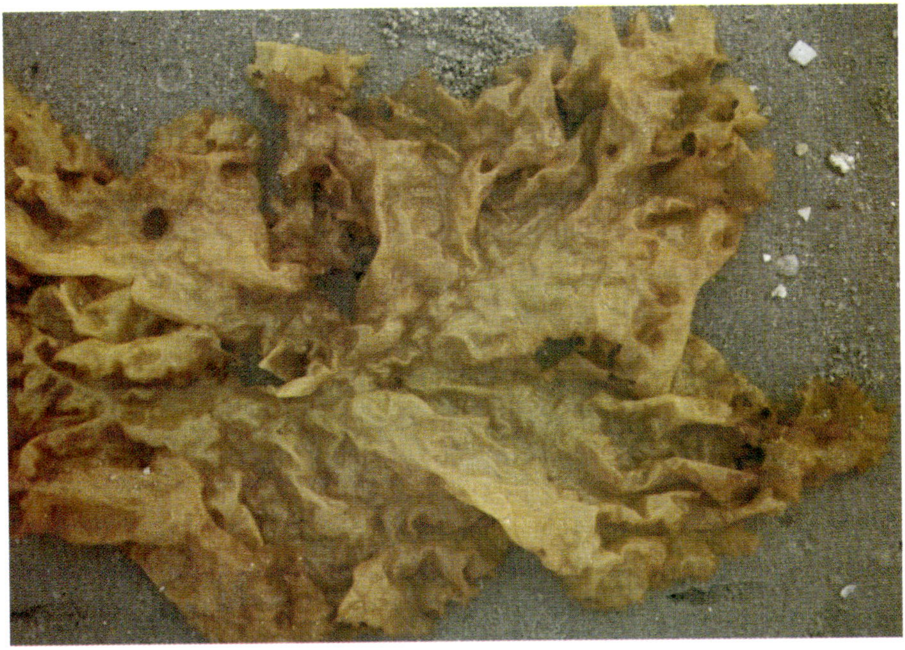

Halymenia dilatata

Halymenia porphyraeformis

Halymenia venusta Børgesen

Plant dark red in color with several flat, flexible frond and slightly lubricous arising from a small holdfast. Blades are variable in shape, subentire or divided into several lobes, margin and surface of blade densely proliferous. Basal attachment by a small disc. They are anchored to various substrata like coral, stone, shells etc. Thallus resembles some forms of *Gigartina* sp., *Iridaea* sp. and *Grateloupia* sp. Dark red fruiting bodies appear as spots and bumps on thallus when in a reproductive state.

Distribution

Gujarat (Dwarka, Saurashtra, Okha), Tamil Nadu (Valinokkam-Killakarai), Andaman and Nicobar Islands, Lakshadweep Island.

Halymenia venusta

Portieria hornemannii (Lyngbye) P. Silva

Thalli are erect, tall, reddish to coralline in colour with flattened branches. Thalli broader at base, branching irregularly, divaricate, pinnate, alternate and subopposite in upper portions. Tips of the branches slightly pointed, curled, margins entire, structure uniaxial, forming arcs.

Distribution

Maharashtra (Bombay), Tamil Nadu (Cape Comorin, Tuticorin, Palk Bay, Krusadai Island, Gulf of Mannar, Mandapam, Tiruchendur), Kerala (Kovalam), Lakshadweep Island (Minicoy), Nicobar Island.

Portieria hornemannii

Amphiroa anceps (Lamarck) Decaisne

Plants red to bright purple in colour, sub-compressed below but flattened above tufted, and 8-10 cm high, fragile, dichotomously branched, sometimes tri- or tetrachotomously branched. Some branches arise from node while some of them below the node. Due to complete calcification of some nodes no articulation is visible. Some articulation is slightly swollen at below ramification. Top joints are fan shaped.

Distribution

Andaman and Nicobar Islands, Gujarat (Okha, Veraval, Saurashtra and Dwarka) Tamil Nadu (Mandapam, Gulf of Mannar, Cape Comorin, Palk-Bay, Tuticorin, Tiruchendur) Maharashtra (Bombay), Kerala and Goa.

Amphiroa fragilissima (Linnaeus) Lamouroux

Thalli are heavily calcified, fragile, thin, whitish pink or purple with repeated dichotomous branching forming extensive mats upto 8 cm thick in shallow tide pools. They are isolated in clumps, common on reef and sheltered location. Thalli rarely found growing with other alga but sometimes with *Enteromorpha sp.* and *Ulva reticulata.*

Distribution

Tamil Nadu (Pamban, Kursadai Island, Gulf of Mannar, Palk Bay, Mandapam, Tuticorin), Maharashtra (Bombay, Malwan), Gujarat Coast, Andhra Pradesh (Visakhapatnam, Gangavaram Port), Goa, Karnataka, Lakshadweep Island (Minicoy), Kerala, Andaman & Nicobar Islands, Indian Coast.

Amphiroa anceps

Amphiroa fragilissima

Corallina berteroi Montagne ex Kützing

Plant thallus are variable in color ranging from pink, purple, red or yellowish. Plants are 10-12 cm in height, calcareous branched, segmented fronds. Thallus has well defined disc-shaped holdfast. Fronds consist of a jointed chain of calcareous segments, each becoming wedge shape. Branches are opposite, resulting in a feather-like appearance. Reproductive organs are urn shaped, usually born at the tips of the fronds. Thallus occasionally found on mollusk shells and few macroalgae.

Distribution

Maharashtra (Bombay), Gujarat (Dwarka).

Jania rubens (Linnaeus) Lamouroux

Plants saxicolous, erect, articulate, 2-5 cm high, densely caespitose, forming closely bunched tufts on the substratum; branching dichotomous; Calcified, rose red segments tightly connected by flexible joints characteristics of this algae. All branches with an erect posture.

Distribution

Gujarat (Dwarka), Maharashtra (Bombay, Malwan), Tamil Nadu (Krusadai Island, Tuticorin, Mandapam, Palk Bay), Andhra Pradesh (Waltair Coast, Visakhapatnam, Gangavaram), Odisha (Chilika Lake), East Coast, Andaman and Nicobar Islands.

Corallina berteroi

Jania rubens

Hypnea musciformis (Wulfen) Lamouroux

Plants are purplish red in colour with 10-22 cm height. Thalli exist in clumps or masses of loosely intertwined, cylindrical branches. Thalli are cartilaginous and highly branched. Branching is variable, irregular, twisted around the axis of other algae. The tips of axis and branches are flattened with broad hooks. Holdfasts are small, inconspicuous or lacking. It can be easily distinguished from other *Hypnea* species by the presence of flattened broad hook at the tips of the branches.

Distribution

Gujarat (Dwarka, Okha, Saurashtra, Veraval, Jalleshwar, Gulf of Kutch), Maharashtra (Bombay, Malwan), Tamil Nadu (Krusadai Island, Tuticorin, Pamban, Mandapam, Porto Novo, Gulf of Mannar, Tiruchendur, Pudumadam), Karnataka (Mangalore), Goa (Anjuna), Lakshadweep Island (Minicoy), Andhra Pradesh (Visakhapatnam), Pondichery, Kerala, Andaman and Nicobar Islands, Bay of Bengal.

Hypnea valentiae (Turner) Montage

Thalli are 10-15 cm in height, have erect branches from the base up to the tips. Branching at the tips giving it a bushy appearance. Thalli have numerous fine lateral branches and few branchlets which are branched occasionally with pointed tips. Several axes with sparse branches, spinous ramuli throughout the length. The satellite structures are longer and firmly fixed to the thallus by their broad base.

Distribution

Maharashtra (Bombay, Malwan), Tamil Nadu (Madras, Krusadai Island, Tuticorin, Mandapam, Muthupet Estuary, Palk Bay, Pamban, Porto Novo, Tiruchendur), Karnataka (Mangalore), Andhra Pradesh (Visakhapatnam), Kerala , Lakshadweep Island (Minicoy), Gujarat (Okha), Pondichery.

Hypnea musciformis

Hypnea valentiae

GIGARTINALES, SEBDENIACEAE

Sebdenia polydactyla (J. Agardh) Parkinson

Plants are robust, dark purple to red in colour. Thallus dichotomously branched. Thallus is 10-12 cm thick at base. Spermatangia aggregated in irregular superficial sori, elongate ovoid. Carpogonia branches are simple. Cystocarp and carposporangia spherical in shape. It occurs in deeper water and detached portions of the plants. It is somewhat slippery and weighty to feel.

Distribution

Gujarat (Okha, Dwarka, Gulf of Kutch), different parts of Indian Coast.

GIGARTINALES, SOLIERIACEAE

Agardhiella subulata (C. Agardh) Kraft and Wynne

Thallus is 30 cm in length, erect and alternate. Holdfast discoid and radially branched. Apices incorporate a limited number of apical cells whose successive derivatives cut off a single periaxial cell in orthostichous rows. Supporting cells of carpogonial branches are homologous to auxiliary cells. They are found in both intertidal and subtidal of Indian Coast.

Distribution

Gujarat (Gulf of Kutch), different parts of Indian Coast.

Sebdenia polydactyla

Agardhiella subulata

Kappaphycus alvarezii Doty

The thalli are smooth, cylindrical, cartilaginous and bushy with multiaxial prostrate or erect branches. Branching not truly opposite, arise often from undersurface of dense thalli. Three different morphotypes are found which are red, green and brown in colour. The plant shows existence of different morphotypes or colour varieties in the cultivation field in response to the environment. Three different colour strains such as deep reddish brown, dark green and yellowish green thalli are found. The thalli are smooth, cylindrical, cartilaginous, bushy, multiaxial, prostrate or erect in habit and consist of irregular indeterminate branches that are rigid. Sometimes, branching is irregularly pinnate, opposite or falsely dichotomous. The indeterminate primary branches arise one at a time from near the tip of non-damaged axis. Secondary branches are common on the older segments. Branches generally curved upward, directed towards the right. Tips of the branches are either pointed or bifurcated. Irregularly arranged, wart-like protuberances are found all along the thalli. Each thallus weighs between 500 gm-1 kg and measures between 20-40 cm in length.

Distribution

Kappaphycus alvarezii was found only in the cultivated sites at Mandapam.

Kappaphycus alvarezii Brown strain

Kappaphycus alvarezii after grazing by fish

Green and Red Strains of *Kappaphycus alvarezii* developed over a period of time in the field.

Kappaphycus alvarezii Green strain

Kappaphycus alvarezii Red strain

Raft cultivation of *Kappaphycus alvarezii*, the author initiated cultivation of this species at Andaman Island. The author suggested for cultivation of *K. alvarezii* in India way back in 1988 and undergone a training for cultivation of this important carrageenophyte at Philippines in 1991.

Cultivation of *Kappaphycus alvarezii* in India

Sarconema filiforme (Sonder) kylin

Thallus is thin and filamentous, subcartilaginous, dichotomously forked, tapering tips. Plants reaching upto 20 cm in height. Thalli internally differentiated into a central medulla and a peripheral cortex with central strands.

Distribution

Lakshadweep Island, Kerala, Gujarat (Gulf of Kutch, Saurashtra, Jalleshwar, Veraval), Tamil Nadu (Tuticorin, Cape Comorin, Mandapam, Tiruchendur), Maharashtra (Bombay), Andhra Pradesh.

Solieria robusta (Greville) Kylin

The thalli are attached to the substratum by a fibrous or discoid holdfast or the hapteron-like structure. The height of the plant is 10-40 cm in length. The colour of the thalli is purple to brown red. The thallus consists of terete or slightly compressed, erect lateral branches which are usually attenuate at their base and sub-acute at their tips. Branching is irregular. Occasionally, 4 or 5 branches arise from a damaged branch and form an umbel.

Distribution

Gujarat (Gulf of Kutch, Okha and Dwarka), Tamil Nadu (Mandapam, Gulf of Mannar, Palk-Bay), Maharashtra (Bombay) Kerala, Bay of Bengal, Lakshadweep Island.

Sarconema filiforme

Solieria robusta

Champia indica Børgesen

Plant thalli are rosy red to darker shades; occurring in dense pyramidal tuft, 15 cm or more in height, thallus conspicuously articulated; segments barrel shaped, 3 mm broad and as long as broad in the main branches; in secondary and other branches much shorter; thallus wall one layer thick, made up of roundish polygonal cells. Vertical filaments transverse the interior of branches through the joints, ramified, ramification reaching the wall; gland cells occur here and there, spherical. Tetraspores scattered over the surface of the thallus, small cystocarps, subspherical in shape and are depressed.

Distribution

Gujarat (Okha, Saurashtra, Gulf of Kutch), Tamil Nadu (Tuticorin).

Champia indica

RHODYMENIALES, CHAMPIACEAE

Champia compressa Harvey

The plants highly branched red in colour up to 3.5 cm in height and 3 mm broad, the branches are terete to slightly compressed irregularly, branched alternately or oppositely, highly mucilaginous, compressed to flattened. A cross section of the thallus shows only one layered wall of more or less squarish cells. Plants form soft clumps on rocks and other seaweeds in the lower intertidal to subtidal zone of shores exposed to moderate wave action.

Distribution

Gujarat (Saurashtra), Maharashtra (Konkan, Malwan), Tamil Nadu, Kerala (Kovalam), South Coast of India, different parts of Indian Coast.

RHODYMENIALES, RHODYMENIACEAE

Botryocladia leptopoda (J. Agardh) Kylin

Plants thallus are purple red, grape like or balloon, tear shaped clustered at top, short stalk, cylindrical, highly branched, attached with holdfast, covered with several swollen vesicles, broader at the upper end. Plants form small groups that arise from single holdfast. It grows on limestone, in crevices in the upper subtidal and tide pools.

Distribution

Maharashtra, Tamil Nadu (Tuticorin, Mandapam, Killakarai, Rameswaram), Gujarat (Gulf of Kutch, Saurashtra, Okha).

Champia compressa

Botryocladia leptopoda

Coelarthrum opuntia (Endlicher) Børgesen.

The plant is purplish- red in colour, beaded in appearance and anchored through a minute and discoid holdfast. Several short, cylindrical and solid stipes arise from the holdfast and bear fastigiated, beaded and jointed fronds. The fronds have well developed apices. The lower segments of the thallus are cunate, ovoid and large whereas the terminal portions are round, small and reveal repeated tripartite branching. The nodes are slender, filiform, hyaline, solid and stalk-like whereas the internodes are hollow and filled with mucilage.

Distribution

Gujarat (Okha), Tamil Nadu (Tuticorin, Gulf of Mannar, Danushkodi, Krusadi Island), different parts of Indian Coast.

Gelidiopsis variabilis (J. Agardh) Schmitz

The height of the frond is 7-10 cm; the primary axis is cylindrical, erect and branched from the bottom. Branches are cylindrical and simple with obtuse apex. Cystocarps are ovoid, sessile on upper branches, either solitary or aggregated. The rounded apical portion of the branches bears tetrasporangia.

Distribution

Tamil Nadu (Tuticorin, Cape Comorin, Mandapam, Tiruchendur), Maharashtra (Bombay), Andhra Pradesh (Visakhapatnam), Gujarat (Jalleshwar, Okha, Saurashtra), Kerala (Kovalam), Southwest Coast of India, Karnataka, different parts of Indian Coast.

Coelarthrum opuntia

Gelidiopsis variabilis

RHODYMENIALES, RHODYMENIACEAE

Rhodymenia dissecta Børgesen

Rhodymenia dissecta shows existence of different morphotypes or colour varieties in the field in response to the environment.Thallus shows different colour strains like reddish brown and yellowish green. Main thalli are 10-15 cm long; and 3-15 mm wide. The thallus is composed of several upright blades, which develop from a common holdfast which is distinctly cuneate at the base. They are branched in palmate manner, often dichotomously. Plant's thallus dichotomously branched up to 5-6 times. The tips of thallus are of acute type. The cystocarpic plants can be easily distinguished by the naked eyes as the cystocarps are scattered on the thallus surface as pink spot. Cystocarpic thallus gave a rough touch to the hand. Cystocarpic plants are found growing in the month of December.

Distribution

Tamil Nadu (Tuticorin, Tiruchendur, Krusadai Island), Gujarat (Okha), Andhra Pradesh (Waltair coast), Lakshadweep Island (Minicoy), South Coast.

CERAMIALES, CERAMIACEAE

Centroceras clavulatum (C. Agardh) Montagne

Thallus is a bushy, dull, purple-red alga. The thallus grows by an apical cell and is about 1-3 cm in height. The filiform thallus attenuates at the apex and repeatedly branches to form fairly regular dichotomy. The bifurcate branches are usually incurved but occasionally out-curved. The axis consists of nodes and internodes. The spines are present in each node and the branch tips are also covered with spines. The rhizoids are colourless, unbranched, traverse, and help in anchorage of the plants to the substratum.

Distribution

Lakshadweep Island, Andaman and Nicobar Islands, Kerala (Kovallum), Gujarat (Dwarka, Jalleshwar, Saurashtra, Veraval), Tamil Nadu (Mandapam, Tiruchendur, Gulf of Mannar, Krusadai Island), Maharashtra (Bombay), Andhra Pradesh (Visakhapatnam), Goa, Karnataka (Mangalore), Odisha (Gopalpur)

Rhodymenia dissecta

Centroceras clavulatum

Ceramium diaphanum (Lightfoot) Roth

The thalli grows on rocky substratum, small pebbles, wooden blocks, ropes and fishing net etc, are predominant in winter and autumn whereas few thalli were found in monsoon season. The thalli are brittle or fragile, purple red in colour and measure upto 9 cm in height. The plants were attached to the substratum by a number of rhizoids. A large number of diatoms and blue green algae were found growing on the thallus surface as an epiphyte.

Distribution

Odisha (Chilika Lake), different parts of Indian Coast.

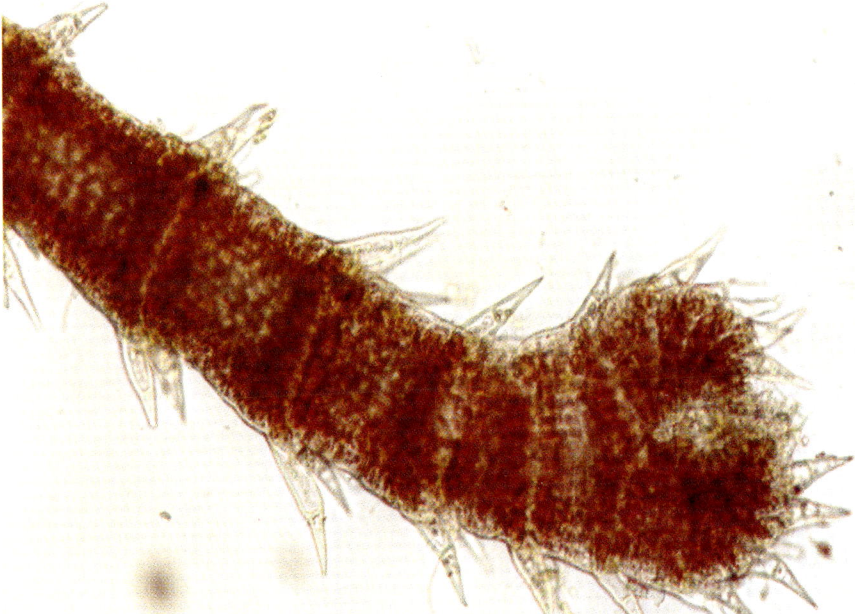

Ceramium diaphanum

Griffithsia corallinoides (Linnaeus) Trevisan

Plants erect, 6-12 cm in height, regularly dichotomously branched fastigiate, branched erect, cells of the upper portions oblong, three times as long as broad; cells in the lower part of the frond subcylindrical, constricted at the junction between superposed cells, three to four times as long as broad; rarely, apex of cells flattened.

Distribution

South-West Coast of India, Different parts of Indian Coast.

Griffithsia corallinoides

Acanthophora spicifera (Vahl) Børgesen

Thalli are purplish brown, brittle, easily broken when fresh, cartilaginous on drying, cylindrical main axis, and branches, growing on subtidal dead coral or rocks. Main axis, with proliferous spinous outgrowths ramuli on determinate branches. These spinous growths are an important feature of the genus. Thalli are attached by holdfast that protrudes many rhizomatous branches.

Distribution

Tamil Nadu (Madras, Mandapam, Krusadai Island, Gulf of Mannar, Palk Bay, Tuticorin), Andhra Pradesh (Visakhapatnam), Gujarat (Jalleshwar, Veraval, Okha, Saurashtra), Goa, Maharashtra, Kerala, Karnataka, Lakshadweep, Andaman and Nicobar Islands.

Acanthophora spicifera

Chondria armata (Kützing) Okamura

Thallus simple or branched terete, attached to rocky substratum by conical holdfast and clumps of rhizoids, branches with pinnately subdivided fronds which are subdistichously arranged. Branches shows determinate growth, 3-8 mm long, cylindrical, tip of the branches bears hair-like structure when young, prominent but rarely persistent when old.

Distribution

Maharashtra (Bombay), Tamil Nadu (Tuticorin, Krusadai Island), Gujarat (Okha, Dwarka, Gulf of Kutch, Saurashtra), Goa, Lakshadweep Island, Andhra Pradesh, Karnataka, Kerala, different parts of Indian Coast.

Chondria armata

Laurencia obtusa (Hudson) Lamouroux

Plants are cartilaginous, fragile, small rarely more than 1.5 cm in high, erect, clavate, purple or red green in color, 6-12 cm in high, cylindrical or uniform diameter. Throughout tapering toward the base into a small disc for attachment, branching alternate to opposite or spiral, shrinking in length from the base to the apex, Cystocarp are conspicuous, distributed on the ramuli and the branches mammilate. Plants are cartilaginous and brittle.

Distribution

Andaman and Nicobar Islands, Maharashtra (Bombay), Tamil Nadu (Krusadai Island, Mandapam, Gulf of Mannar, Palk Bay), Gujarat, (Dwarka, Saurashtra), Goa, Andhra Pradesh, Karnataka.

Laurencia pedicularioides Børgesen

The fronds are erect, nearly up to 10 cm high and possess several terete axes densely tufted below with entangled and somewhat coalescing basal branches. The thalli are firmly anchored to the substratum by a basal adhesive holdfast or disc from which arise the axis bearing several branches. The branching is alternate, opposite or verticillate. The fronds are generally pinkish-brown but usually show a pink hue at the apices.

Distribution

Lakshadweep Island, Maharashtra (Bombay), Tamil Nadu, Gujarat (Dwarka, Saurashtra, Gulf of Kutch, Okha).

Laurencia obtusa

Laurencia pedicularioides

Polysiphonia substilissima Montagne

The thallus is purple brown in colour; slimy, soft and grows upto 10 cm long. The main axis grows as a prostrate system which gives rise to the secondary branches as well as the rhizoids. The rhizoids are colourless and arise from the parietal cell. The secondary branches are narrow, thin and are long. The branching is subdichotomous and alternate. The filament is made up of 4 pericentral cells with a siphon in the centre. The pericentral cells are tetragonal in shape and are covered by a gelatinous layer.

Distribution

Odisha (Chilika lake), Tamil Nadu (Pulicat lake).

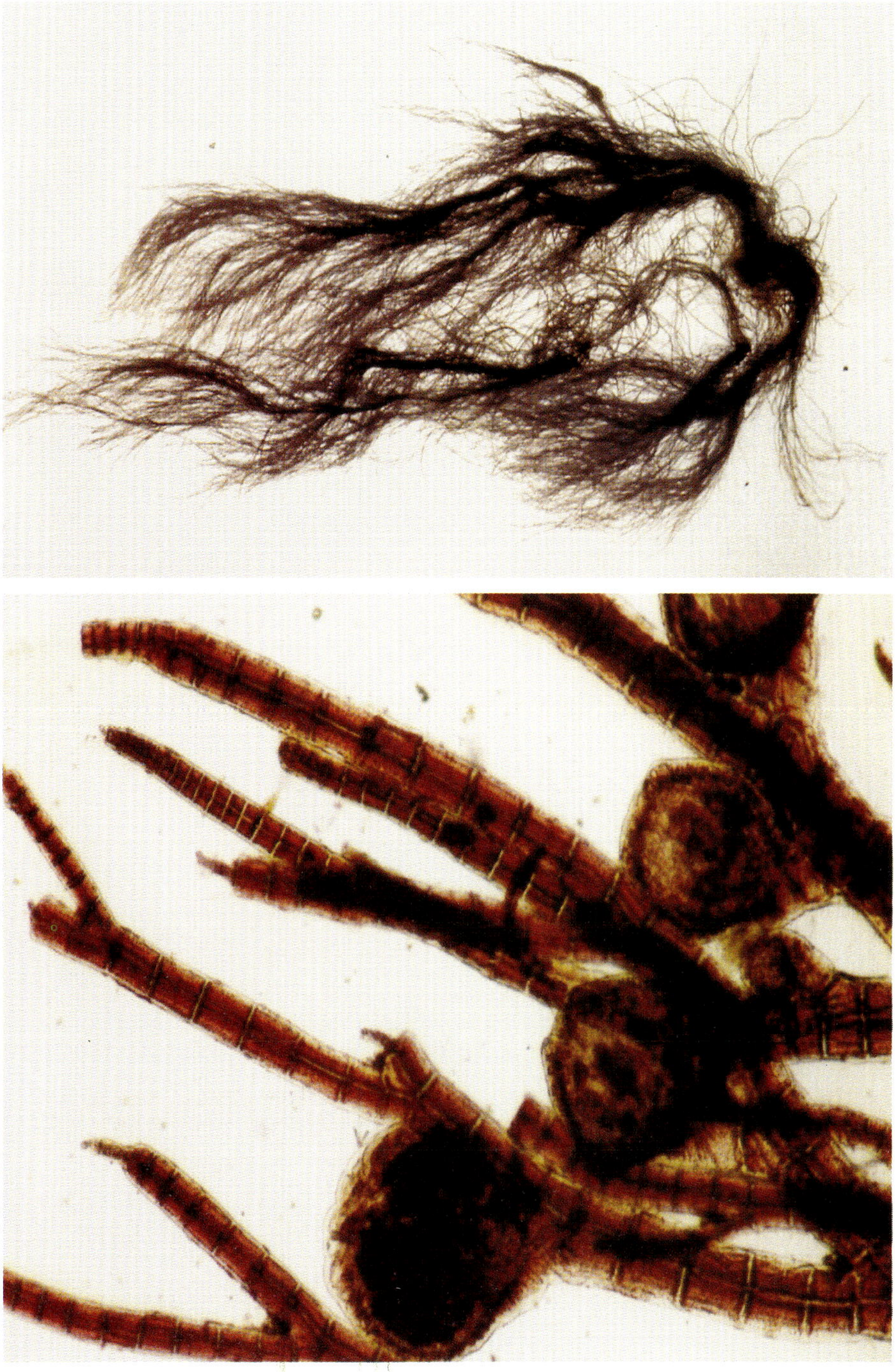

Polysiphonia substilissima

References

Abbott, A.I. (1995). *Taxonomy of economic seaweeds*, Vol. V, Publication of California Sea Grant college system. pp. 254.

Baweja, P. (2001). Developmental and histochemical studies on *Grateloupia filicina* (Lamouroux) C. Agardh- An important carrageenophyte. Thesis, M. Phil. University of Delhi, Delhi. pp. XXIX + 50. (Under the supervision of Dr. Dinabandhu Sahoo).

Baweja, P. (2005). Structure and reproduction of *Grateloupia lithophila* (Rhodophyta) and its Evaluation for commercial exploitation with *Grateloupia filicina* as a source of food and carrageenan. Ph.D. Thesis, University of Delhi. pp. XLIII-148 (Under the supervision of Dr. Dinabandhu Sahoo).

Calumpong, H.P. and Menez, E.G. (1996). *Field guide to the common mangroves, seagrasses and algae of the Phillippines.* Bookmark, Inc. 264-A Pablo Ocampo Sr. Ave. Makati City, Philippines. pp. 197.

Desikachary. T.V., Krishnamurthy, V. and Balakrishnan, M.S. (1998). *Rhodophyta* Vol. II-B Madras Science Foundation, Chennai. pp. 359. XXX Fig. 89.

Fritsch, F.E. (1935). *The structure and reproduction of the algae Vol. I.* Vikas Publishing House Pvt. Ltd., New Delhi. pp. 939.

Fritsch, F.E. (1945). *The structure and reproduction of the algae Vol. II.* Vikas Publishing House Pvt. Ltd., New Delhi. pp. 791.

Isao. T., Nang, H.Q., Dhin, N.H., Shogo, A. and Tadao, Y. (2005). *The common marine plants of Southern Vietnam.* Japan Seaweed Association, Kochi, Japan. pp. 250.

Krishnamurthy, V. (2000). *Algae of India and neighbouring countries I. Chlorophycota*, Oxford & IBH Publishing Co. Pvt. Ltd. New Delhi. pp. 210.

Littler, D.S., Littler, M.M., Bucher, K.E. and Norris, J.N. (1989). *Marine plants of the Caribbean: A field guide from Florida to Brazil.* Smithsonian Institution Press, Washington, D.C. pp. 263.

Magruder, W.H. and Hunt, J.W. (1979). *Seaweeds of Hawaii.* The Oriental Publishing Company, Honolulu, Hawaii. pp. 116.

Sahoo Dinabandhu (1985). Histochemical and developmental studies on phases of reproduction in red algae-*Centroceros clavatum* (C. Agardh) Montagne. M. Phil. Thesis, University of Delhi, Delhi, India. pp. XXIII+35.

Sahoo Dinabandhu (1988). Histochemical and ultrastructural studies on *Cystseira indica* from Port Okha. Thesis, Ph.D. University of Delhi, Delhi, India. LXVI-94.

Sahoo Dinabandhu (2000). "*Farming the Ocean: Seaweeds Cultivation and Utilization*". Aravali Books International, New Delhi, India. pp. XI+44.

Sahoo Debasish, (2005). Assessment of diversity of seaweeds of Chilika Lake and their evaluation for commercial exploitation. Thesis, Ph. D. University of Delhi, Delhi, India. pp. CXCV + 55. (Under the supervision of Dr. Dinabandhu Sahoo).

Sahoo Dinabandhu., Nivedita, and Debasish. (2001). "*Seaweeds of Indian Coast*" APH Publishing Corporation, New Delhi, India. pp. XXI + 283.

Sahu, N. (2004). Studies on *Gracilariods* from Chilika Lake- structure development and Agar quality. Thesis, Ph.D. University of Delhi, Delhi, India. pp. LIV + 132. (Under the supervision of Dr. Dinabandhu Sahoo).

Savindra Kumar (2007). Developmental and Histochemical Studies in Red Alga *Rhodymenia dissecta* Bøergesen. Thesis, M. Phil. University of Delhi, Delhi. pp. XX-72. (Under the supervision of Dr. Dinabandhu Sahoo).

Silva, P.C., Basson, P.W. and Moe, R.L. (1996). *Catalogue of the benthic marine algae of the Indian Ocean.* University of California Press. Berkely. pp. 1259.

Singh, M.A. (2007). Study of seaweeds diversity during the premonsoon season at Mandapam coast. Thesis, M. Phil. University of Delhi, Delhi. pp. XXVII + 68 (Under the supervision of Dr. Dinabandhu Sahoo).

Singh, S. (2005). Light and electron microscopic studies in four tropical *Porphyra* species. pp. LIX-111. (Under the supervision of Dr. Dinabandhu Sahoo).

Srinivasan, K.S. (1969). *Phycologia Indica (Icones of Indian Marine Algae) I.* Botanical Survey of India, Calcutta, pp. 52.

Srinivasan, K.S. (1973). *Phycologia Indica (Icones of Indian Marine Algae) II.* Botanical Survey of India, Calcutta, pp. 60.

Tokuda, H., Kawashima, S., Ohno, M. and Ogawa, H. (1994). *A photographic Guide : Seaweeds of Japan.* Ikebukuro Nishiguchi sky bldg, Tokyo, Japan. pp. 194.

Tseng, C.K. (1983). *Common seaweeds of China*, Science Press, Bejing, China, pp. 316.

Index